Art and Complexity

Art and Complexity

Editors

John Casti
Santa Fe, New Mexico, USA

Anders Karlqvist
Stockholm, Sweden

2003

ELSEVIER
Amsterdam · Boston · London · New York · Oxford · Paris
San Diego · San Francisco · Singapore · Sydney · Tokyo

ELSEVIER SCIENCE B.V.
Sara Burgerhartstraat 25
P.O. Box 211, 1000 AE Amsterdam, The Netherlands

First edition 2003

Library of Congress Cataloging in Publication Data
A catalog record from the Library of Congress has been applied for.

British Library Cataloguing in Publication Data
A catalogue record from the British Library has been applied for.

ISBN: 0-444-50944-5

♾ The paper used in this publication meets the requirements of ANSI/NISO Z39.48-1992 (Permanence of Paper).
Printed in The Netherlands.

Cover illustration: © Jackson Pollock, Waterbirds, 2002, c/o Beeldrecht, Hoofddorp

Contributors List

John D. Barrow, *DAMTP, Centre for Mathematical Sciences, Cambridge University, Wilberforce Road, Cambridge CB3 0WA, United Kingdom; e-mail: J.D.Barrow@damtp.cam.ac.uk*

John Casti, *1399 Hyde Park Road, Santa Fe, NM 87505 USA; e-mail: Castiwien@cs.com*

James P. Crutchfield, *Santa Fe Institute, 1399 Hyde Park Road, Santa Fe, NM 87505 USA; e-mail: chaos@cs.com*

Murray Gell-Mann, *Santa Fe Institute, 1399 Hyde Park Road, Santa Fe, NM 87505 USA; e-mail: gell-mann@cs.com*

Bobo Hjort, *Landskapsplanering, SLU, Box 7012, SE-750 07 Uppsala, Sweden; e-mail: bobo.hjort@lpul.slu.se*

Anders Karlqvist, *Swedish Polar Research, Box 50005, SE 104 05 Stockholm, Sweden; e-mail: anders.karlqvist@polar.se*

Laurent Mignonneau, *IAMAS Institute of Advanced Media Arts and Sciences, 3-95 Ryoke-cho, Ogaki-shi, Gifu 503-0014, Japan; e-mail: laurent@mic.atr.co.jp*

Tor Nørretranders, *Tuborg Parkvej 2.4.2, DK-2900 Hellerup, Denmark; e-mail: tor@tor.dk*

Tim Perkis, *1050 Santa Fe Ave., Albany, CA 94706; e-mail: perkis@artifact co*

Christa Sommerer, *IAMAS Institute of Advanced Media Arts and Sciences, 3-95 Ryoke-cho, Ogaki-shi, Gifu 503-0014, Japan; e-mail: christa@iamas.ac.jp*

Marcia Southwick, *1399 Hyde Park Road, Santa Fe, NM 87501; e-mail: SMouthwick@aol.com*

Richard Taylor, *Department of Physics, University of Oregon, Eugene, OR 97403-1274; e-mail: rpt@darkwing.uoregon.edu*

Gail Wight, *2716 Martin Luther King Jr. Way, Berkeley, CA 94703 e-mail: gailw@notochord.org*

Contents

Preface

The concept of "complexity" is a lot like other informal—but very useful—everyday notions like "truth," "beauty," "love," and "justice," residing as much in the eye of the beholder as in some objective, platonic realm beyond space and time. But there is an emerging *science* of complex systems, which is often termed the "science of complexity," suggesting a formalization of what we mean by "complexity" in objective, mathematical, and computational terms. So it's of more than passing interest to investigate whether a formal notion of complexity accords with what practitioners of the more humane arts, such as literature, painting, poetry, and music mean by a work of art being *complex*.

With this exploration of the interface between complexity and art as the leitmotif, the Swedish research agency, FRN, sponsored a one-week workshop to bring together complexity scientists interested in art and artists interested in complexity to exchange views of the issue of how complexity and art fit together. This meeting was held in May 1998 at the scientific station of the Royal Swedish Academy of Sciences in Abisko, Sweden, a small village in Lapland, nearly 100 miles north of the Arctic Circle. The pristine beauty and stark surroundings of this remote location, as well as the warm collegiality engendered by a small group living together on a 24-hours-per-day basis for a week, contributed to many stimulating conversations and contemplations on the relationship between art and complexity. This book is a pale attempt to capture the gist of these deliberations.

Within the pages that follow, the reader will see an impressive array of explorations of the art/complexity interface, including discussions of whether "good" art is "complex" art, how artists see the term "complex," what poets try to convey in words about complex behavior in nature, and a whole lot more. Taken as a whole, this volume serves as an ongoing testament to the ability of artists and scientists to communicate to unravel mysteries in the world around us.

John Casti
Santa Fe and Vienna

Anders Karlqvist
Stockholm

February 2002

Art and Complexity
J. Casti and A. Karlqvist (editors)

Art and Science—Les Liaisons Dangereuses?

John D. Barrow

"We are reluctant, with regard to music and art, to examine our sources of pleasure and strength. In part we fear success itself—we fear that understanding might spoil enjoyment. Rightly so! Art so often loses power when its psychological roots are exposed"

—Marvin Minsky

1 WHY LOOK AT ARTS AND SCIENCE?

Most artists are very nervous of scientific analysis. They feel it destroys something about the human aspect of creativity. The fear (possibly real) of unsubtle reductionism—music is nothing but the trace of an air pressure curve—is widespread. As a corollary, one finds the equally pernicious view that science has nothing to offer the arts, that they must transcend all attempts to capture them. Abraham Moles, in his classic book *Information Theory and Aesthetic Perception* [15], writes of the view that

> "Aesthetic information cannot be translated. It does not draw on a universal repertoire of knowledge that the sender and receiver have in common."

Indeed, some fear that too much analysis will only break the spell. Likewise, most scientists see the creative arts as an entirely subjective development that long ago left science to tread the long road to objective truth alone. Whole books have been written about this bifurcation, but here I want to talk about some interesting points of contact between art and science that are facilitated by the growth in our understanding of complex organization and pattern. I want to explore a (British) compromise position that sees a useful reciprocal relationship between parts of the arts and sciences. I believe that the sciences of complexity have a lot to learn from the creative arts. The arts display some of the most intricate known examples of organised complexity. Likewise, the creative arts may have something to learn from an appreciation of what complexity is and how it comes about [6]. E.O. Wilson, in his recent book, *Consilience* [24], remarks that

> "The love of complexity without reductionism makes art; the love of complexity with reductionism makes science."

In section 2 we are going to review an old and little known attempt to quantify aesthetic complexity that was made in 1933 by the mathematician George Birkhoff. In section 3 we take a look at landscape art from a different perspective; and in section 4 consider whether criticality is a useful concept in musical appreciation.

2 THE GEOMETERS' APPROACH

There have been many minor forays into this area. They all face a problem of under determination that comes of regarding artistic composition as a process with some number of dimensions of time (T) and space (S), so that a line of print or a frieze is an example of a one-dimensional spatial process (S); speech or music is a one-dimensional temporal process (T); painting or photography is therefore $S \times S$; a sound tape or knitting is $S \times T$; sculpture of architecture is $S \times S \times S$; movies are $S \times S \times T$; theatre or opera are $S \times S \times S \times T$; perhaps a movie with fractal images might even be $S^d \times T$ where d is not an integer. But while this might be useful for classification purposes it sheds no light on the real thing. Nonetheless it might well be instructive to focus upon particular well-defined artistic creations that exist within a particular medium. While it makes little sense to compare aesthetic effects from medium to medium (is that painting nicer than that symphony?!) we might make some progress in isolating what it is about complexity (or simplicity) that affects us in a single creative activity that is not too complicated. The most extensive attempt to do this was made by Garrett Birkhoff in the 1930s and it is recounted in great detail in his book *Aesthetic Measure* [8]. Birkhoff attempted to capture something of the "unity in diversity" that is engaging about art and natural beauty by means of a simple formula. Defining **O** to be some measure of inherent *Order* and **C** to be some measure of *Complexity* he then defined a quotient formula for *Aesthetic Measure* (**M**) to be applied within some class of similar artistic creations:

$$\text{Aesthetic Measure} = \text{Order}/\text{Complexity}$$

He pursued a long-term project to apply this simple measure to all sorts of creative works of art. The amount of information that it uses depends entirely upon how one defines **O** and **C**. Birkhoff adopts fairly simple geometrical measures for these indicators. Today, he would probably appeal to the concept of self-similarity or fractal dimension to capture more of what it is that we find visually or aurally appealing. But, let's see what he did.

To get a feel for Birkhoff's approach in the simplest possible situation consider his evaluation of the Aesthetic Measure of polygons. He defines their Complexity, **C**, by the number of indefinitely extended straight lines which contain all of a polygon's sides. He defines their Order, **O**, by the expression

$$\mathbf{O} = \mathbf{V} + \mathbf{E} + \mathbf{R} + \mathbf{HV} - \mathbf{F}$$

where **V** is a measure of vertical symmetry (equal to +1 is there is a vertical symmetry axis and 0 otherwise). The Equilibrium Element, **E**, measures whether there is optical (rather than mechanical) equilibrium of the figure. This is judged to exist if the centre of the figure's area lies between vertical lines through extreme points and is at a distance from them greater than one sixth of the horizontal width of the polygon. If this holds $\mathbf{E} = +1$. If not, but the polygon is in mechanical equilibrium then $\mathbf{E} = 0$, otherwise $\mathbf{E} = -1$; in particular, $\mathbf{E} = 1$ if $\mathbf{V} = 1$. Rotational Symmetry is measured by **R**, with $\mathbf{R} = +1$ if there exists central symmetry (polygon unchanged by rotation through 180 degrees). In general, if it is invariant under a rotation angle of $360°/Q$ then it scores a value of $\mathbf{R} = 1/2Q$ for values of $\mathbf{Q} < 6$ and $\mathbf{R} = 3$ for $\mathbf{Q} > 6$.

The quantity **HV** identifies the presence of a horizontal-vertical network which is pleasing to the eye. If a polygon has all its sides along a uniform network of vertical and horizontal lines (like a square with horizontal base or a Swiss cross, as in figure 1) then it scores $\mathbf{HV} = +2$, because there exist two independent (horizontal and vertical) translations under which the network is invariant. If all the sides lie on two sets of parallel lines equally inclined to the vertical and fill a diamond shape then $\mathbf{HV} = +1$.

The last contribution to **O** is **F**, the Unsatisfactory Form Factor, and it enters with a negative sign. Its value is defined to be $\mathbf{F} = 0$, if any of the following hold:

1. The inter-vertex distances are too small ($< 0.1\times$ width of the polygon)
2. Angles between non-parallel sides too small ($< 20°$)
3. A shift of vertices by < 0.1 of distance to the nearest vertex introduces a new element of Order in $\mathbf{V}, \mathbf{R}$, or **HV**.
4. There is an unsupported side that re-enters the figure.
5. There is more than one type of niche in the sides of the figure.
6. Sides point in directions other than vertical or horizontal
7. $\mathbf{V} = \mathbf{R} = 0$

If all but one of these seven conditions hold then $\mathbf{F} = +1$, otherwise $\mathbf{F} = +2$.

As an example of the evaluation of **O** and **C**, consider the polygon defined by the Swiss cross in figure 1.

The Complexity is given by $\mathbf{C} = 8$ since we have to draw four horizontal and four vertical lines in order to pass through all the sides of the polygon. The Order is given by combining the values of $\mathbf{V} = 1$, $\mathbf{E} = 1$, $R = 2$ (since the cross is invariant under rotation by 90°, so $\mathbf{Q} = 4$), $\mathbf{HV} = 2$, and $\mathbf{F} = 0$, to give $\mathbf{O} = 6$; hence the Aesthetic Measure is $\mathbf{M} = 3/4$.

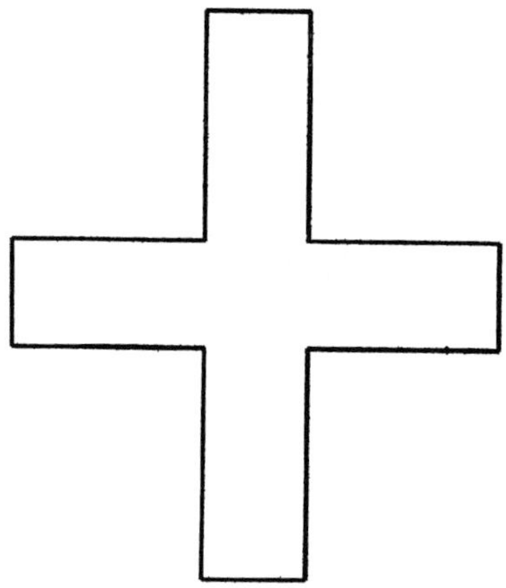

FIGURE 1 A polygonal figure with $\mathbf{C} = 8$ and $\mathbf{O} = 6$.

There are a few generalities which are evident from Birkhoff's definitions. All triangles must have one of the set of six values

$$\mathbf{M}(\text{triangles}) = \{7/6, 2/3, 0, -1/3, -2/3, -1\}.$$

Since $\mathbf{V} < 1$, $\mathbf{E} < 1$, $\mathbf{R} < 3$ and $\mathbf{HV} < 2$, for polygons we have the bounds

$$\mathbf{O}(\text{polygons}) \leq 7$$

$$\mathbf{M} < 7/\mathbf{C}.$$

What is the largest value of $\mathbf{M}$ for these shapes? For a square with a horizontal side, we have simply,

$$\mathbf{O}(\text{square}) = 6,\ \mathbf{C}(\text{square}) = 4,\ \mathbf{M}(\text{square}) = 1.5$$

Now $\mathbf{M}$ for any other polygon can only exceed or equal the value for the square if $\mathbf{C} = 3$ or 4. But $\mathbf{C} = 3$ means we have triangles and $\mathbf{M}(\text{triangles}) \leq 7/6 < \mathbf{M}(\text{square})$; the remaining possibility is $\mathbf{C} = 4$. Now $\mathbf{HV} \neq 0$ only if we have a square ($\mathbf{M} = 1.5$), a rectangle ($\mathbf{M} = 1.25$) or a diamond ($\mathbf{M} = 0$); otherwise $\mathbf{HV} = 0$ and $\mathbf{R} = \mathbf{V} = 0$, so $\mathbf{M} < 2/4 = 0.5$. Thus the value of $\mathbf{M} = 1.5$ obtained for the square is the largest possible for any polygon.

In figure 2 a collection of 90 different polygonal shapes are shown along with the value of their Aesthetic Measure, $\mathbf{M}$, in descending order.

There are many other set of aesthetic creations to which Birkhoff applied his basic idea. I want to highlight just one of them. It is the next level in complexity that one might imagine after polygonal structures and tilings which employ only straight edges. It is the form of a vase, viewed in projection so that one can treat

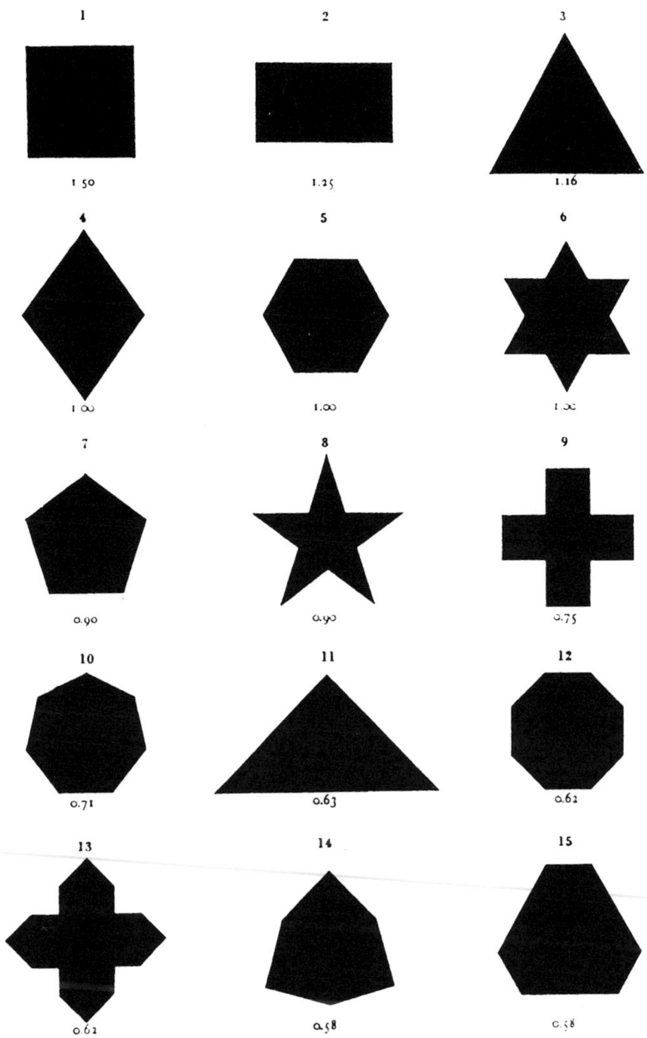

FIGURE 2 A collection of 90 polygonal shapes and the values of their aesthetic measures, M. (Reprinted by permission of the publisher from AESTHETIC MEASURE by George D. Birkhoff, Cambridge, Mass.: Harvard University Press, Copyright © 1933 by the President and Fellows of Harvard College.)

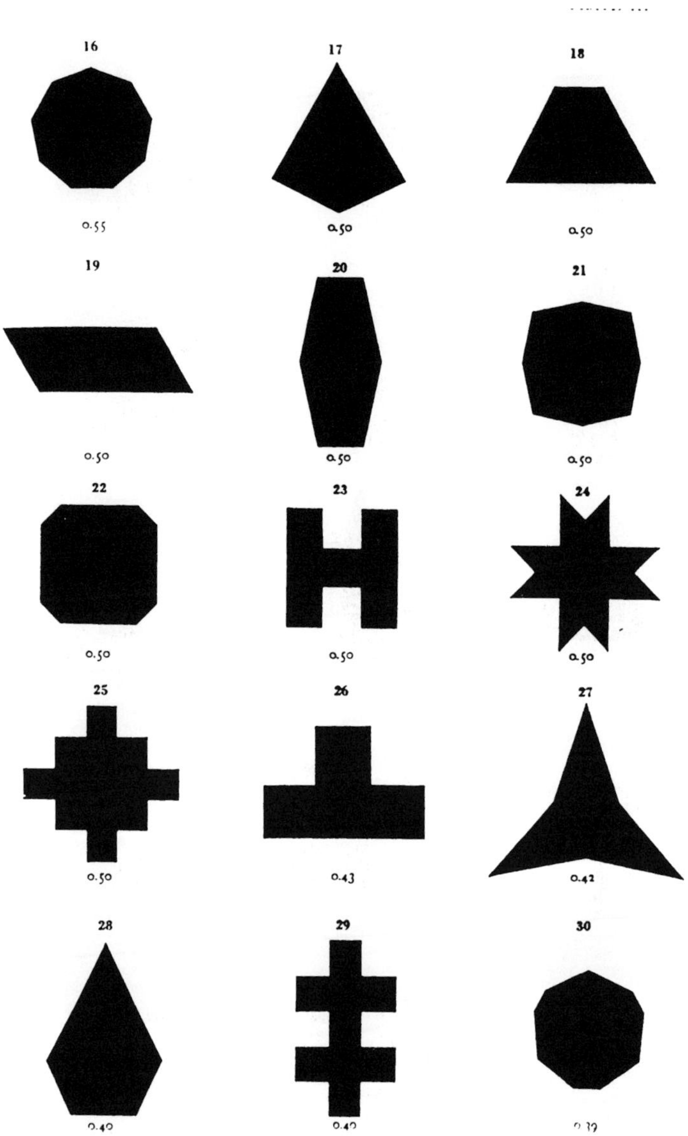

FIGURE 2 Continued.

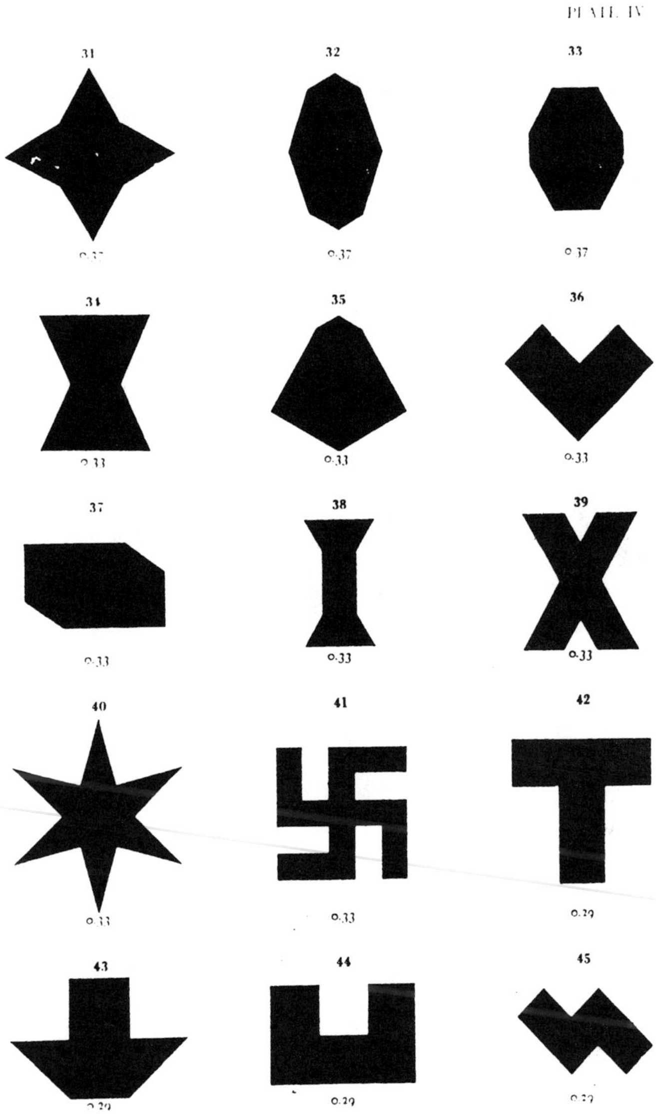

FIGURE 2 Continued.

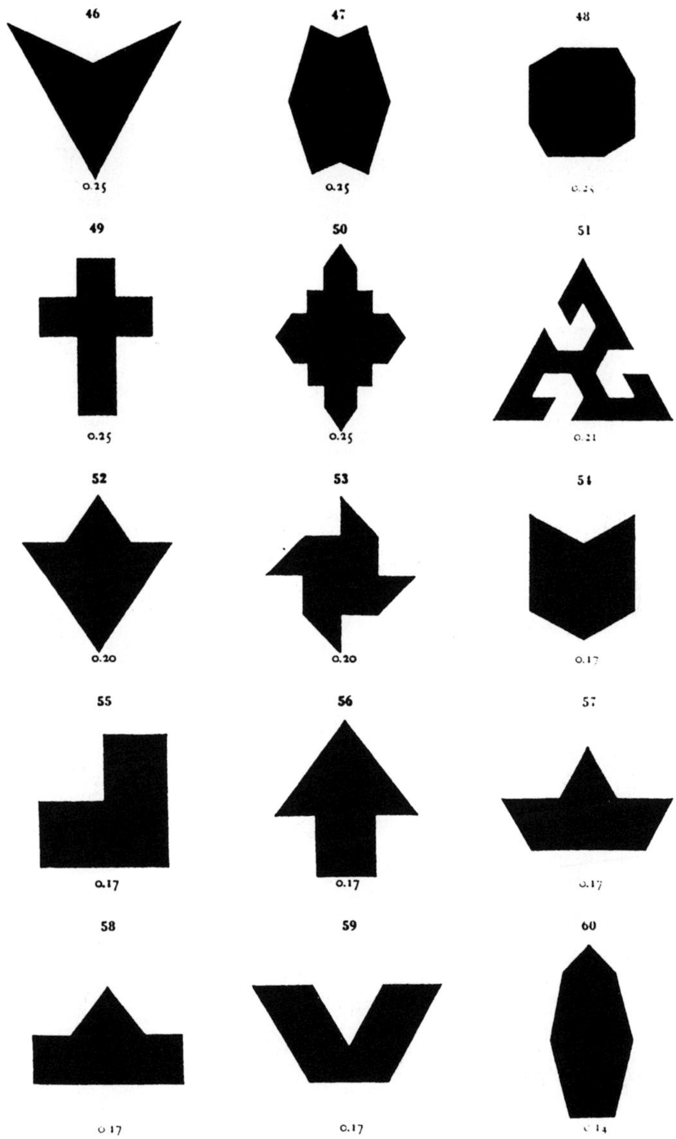

FIGURE 2 Continued.

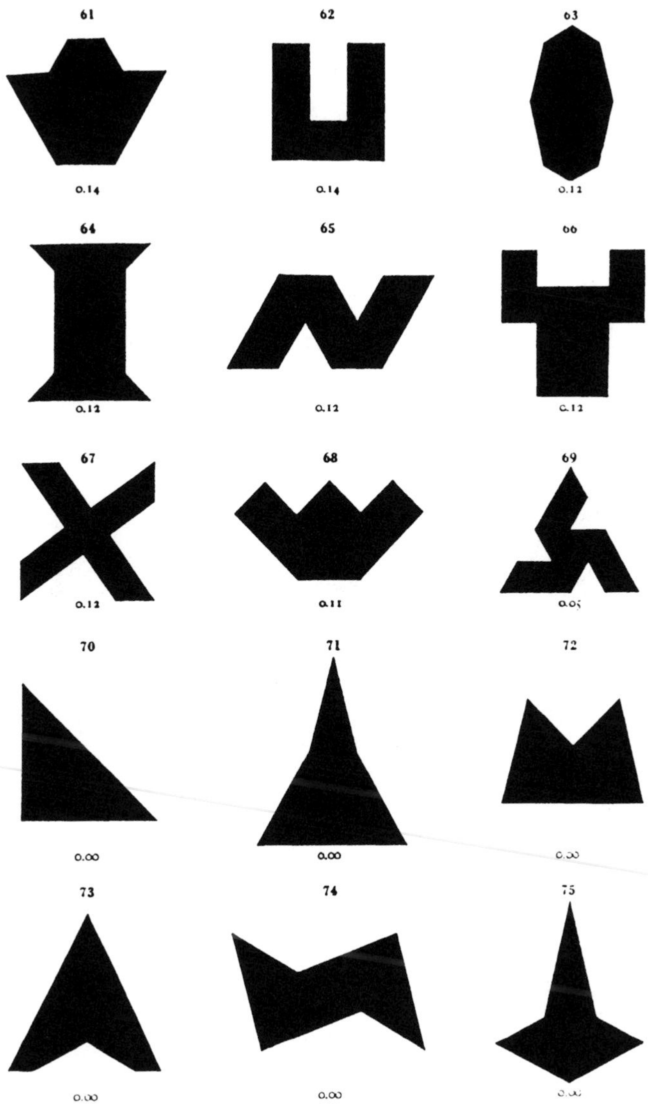

FIGURE 2 Continued.

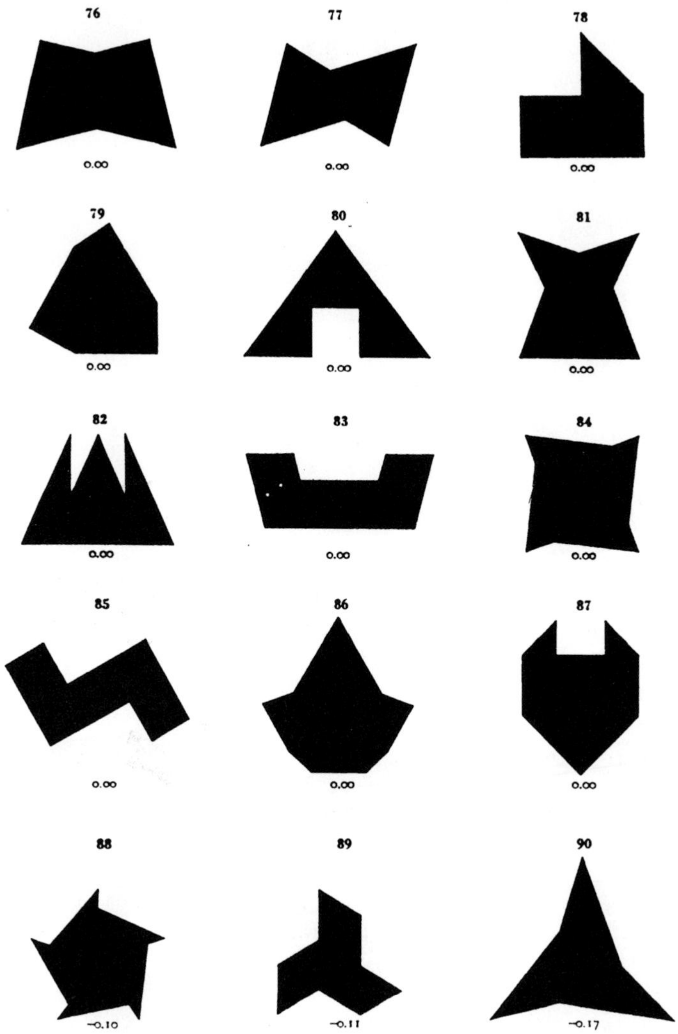

FIGURE 2 Continued.

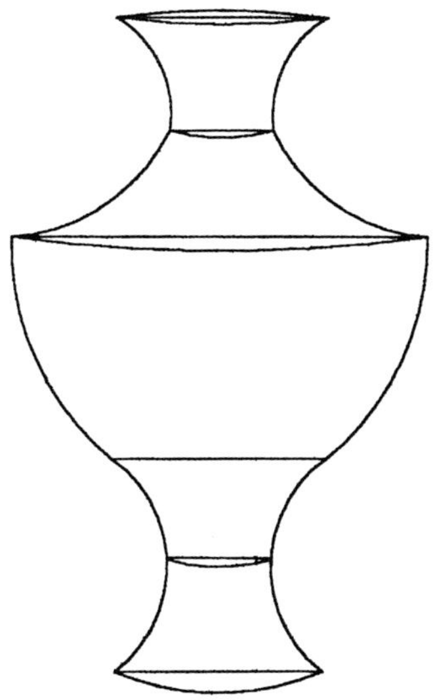

FIGURE 3 A typical classical vase contour.

it as a two-dimensional shape. Colourings will be ignored as a first approximation. Besides being an interesting application of Birkhoff's naïve Measure, it also highlights a nice simply family of aesthetic patterns which might be interesting subjects for more sophisticated quantitative measures.

The basic vase form is shown outlined in figure 3. Lateral symmetry is assumed always to be present. Likewise, there are always two curvilinear sides and two convex elliptical ends.

Our aesthetic evaluation of the contour of this form is influenced by several simple geometrical features of the outline: places where the contour ends (the lip and base of the vase), places where the tangent to the contour is vertical, places where the tangent changes abruptly (corner points) and points of inflexion in the tangent direction. These special points on the contour form our aesthetic impression of its form. They are indicated in figure 4.

Birkhoff defined the *Complexity*, **C**, of a vase to be equal to the number of special points where the tangent to the vase contour is vertical, has inflexions, corner points or end points. By inspection, it is clear that

$$6 < \mathbf{C}(vase) < 20.$$

The Order of the vase will be defined by

$$\mathbf{O} = \mathbf{H} + \mathbf{V} + \mathbf{HV} + \mathbf{T}$$

where $\mathbf{H} < 4$ is the number of independent horizontal distance relations that are in the ratio 1 : 1 or 2 : 1; $\mathbf{V} < 4$ the number of independent vertical distance

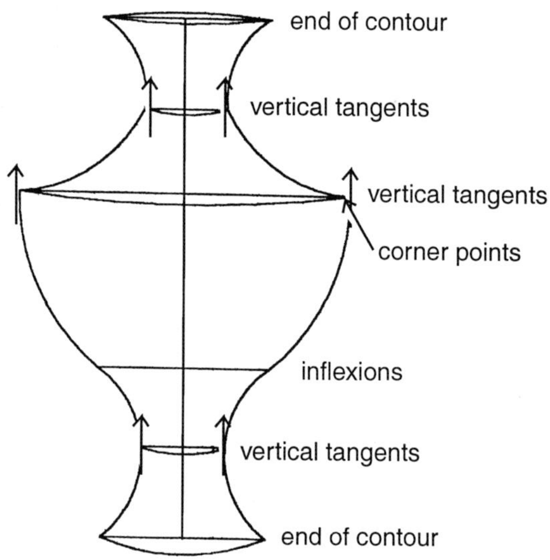

FIGURE 4 A vase contour with characteristic visual points indicated. Not all of these points need occur in a profile but those that do will dominate the visual appraisal.

relations that are in the ratio 1 : 1 or 2 : 1; **HV** < 2 the number of independent inter relations between vertical and horizontal distances that are in the ratio 1 : 1 or 2 : 1, and **T** < 4 is the number of independent perpendicular and parallel relations between tangents plus the number of vertical tangents at end points and inflexions and the number of characteristic tangents through an adjacent centre.

Some applications of this system of rules to evaluate **M** for four classic Ming, Sung and T'ang Chinese vase forms are shown in figure 5. Three experimental vase forms, generated by Birkhoff so as to create profiles with high values of **M**, are shown in figure 6 for comparison. Note that the artificial forms have much greater values of **M** than the real vase profiles.

The challenge presented by these old investigations is to generalise or improve these measures of aesthetic appeal to include what we have learned about fractality and self-similarity, introduce three-dimensional perspective and modern computer display to explore the limits and scope of particular quantifications of shape and harmony in line and contour. For some discussion of other quantitative measures of aesthetic content see Eigen and Winkler [11]; for different discussion of the subject see the book by McAllister [16].

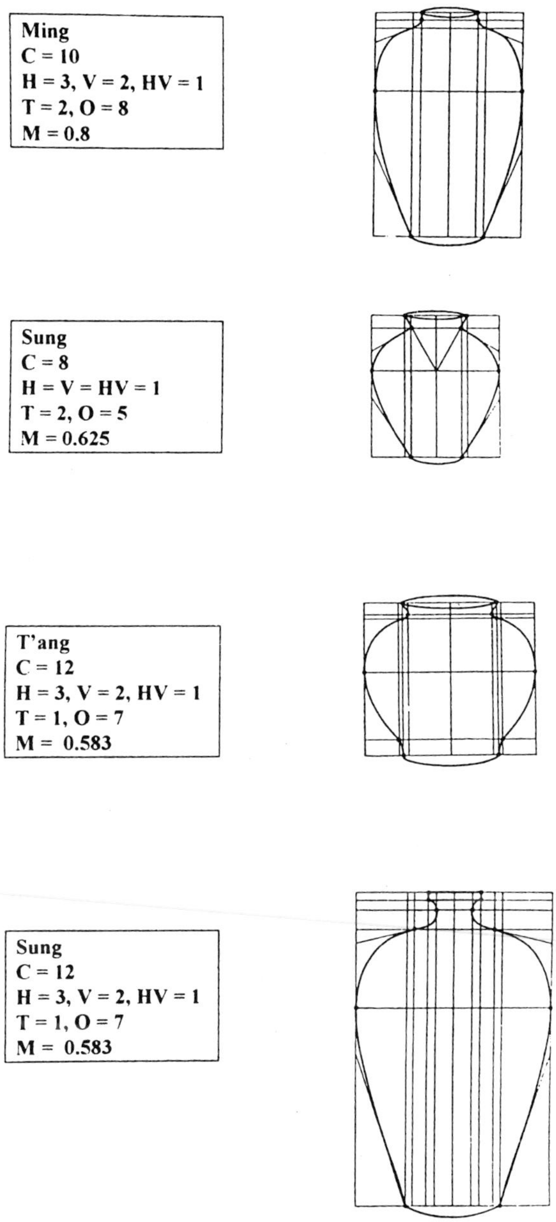

FIGURE 5 Analysis of **O**, **C**, and **M** values for four classical Chinese vase forms according to the aesthetic formula defined in the text. (Reprinted by permission of the publisher from AESTHETIC MEASURE by George D. Birkhoff, Cambridge, Mass.: Harvard University Press, Copyright © 1933 by the President and Fellows of Harvard College.)

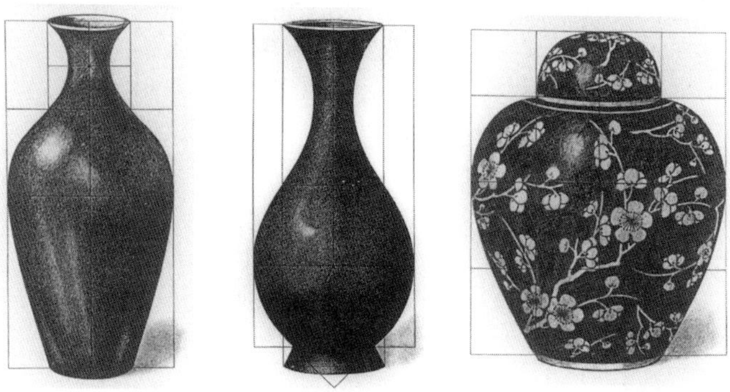

(a) (b) (c)

FIGURE 6 Three vase forms with high aesthetic measures artificially generated by Birkhoff: (a) $\mathbf{C} = 10, \mathbf{H} = 4, \mathbf{V} = 3, \mathbf{HV} = 1, \mathbf{T} = 2, O = 10$, and $\mathbf{M} = 10/10 = 1$. (b) $\mathbf{C} = 12, \mathbf{H} = 4, \mathbf{V} = 3, \mathbf{HV} = 2, \mathbf{T} = 4, \mathbf{O} = 13$, and $\mathbf{M} = 13/12 = 1.08$ (c) $\mathbf{C} = 7, \mathbf{H} = 2, \mathbf{V} = 1, \mathbf{HV} = 2, \mathbf{T} = 2, \mathbf{O} = 7$, and $\mathbf{M} = 7/7 = 1$. (Reprinted by permission of the publisher from AESTHETIC MEASURE by George D. Birkhoff, Cambridge, Mass.: Harvard University Press, Copyright © 1933 by the President and Fellows of Harvard College.)

3 THE EVOLUTIONARY PSYCHOLOGISTS' APPROACH

Let us turn now to a completely different perspective on aesthetic appreciation. Several speakers at this meeting have raised the question "what pictures do we like?" John Casti challenged us to grade a collection of pieces of computer art according to their aesthetic appeal. The Aesthetic Measure of Birkhoff was a rather rigid way of capturing aspects of the symmetry and cleanness of line in plane figures that are appealing to us. However, there is an entirely different evolutionary perspective that can be brought to bear upon these aesthetic questions. We can ask whether any of our aesthetic preferences make sense as by-products of adaptations which enhanced the survival probability of our ancestors 0.5–2 million years ago [6, 5, 23]. For some arts (e.g., story telling, sculpture, dance) we can detect an adaptive activity from which they derive. One of the most striking applications of this approach is to the appreciation of landscape and landscape art. If we take ourselves back to the type of ancient environment in which our African savannah-dwelling ancestors spent the bulk of their evolutionary history we can argue that a sensitivity (either attractive or repulsive) for certain types of environment would have definite survival value and would therefore be more likely to be inherited in the long run [1, 2, 6, 7, 12, 18, 19]. What might we deduce from this psychobiological approach to landscape appreciation? First, that landscapes in which we can see without being seen should be appealing. Such landscapes, which provide us with the ability to see without being seen are typified by savannah landscapes, are typical of our ornamental parks and gardens (see figure 7). They offer both prospect

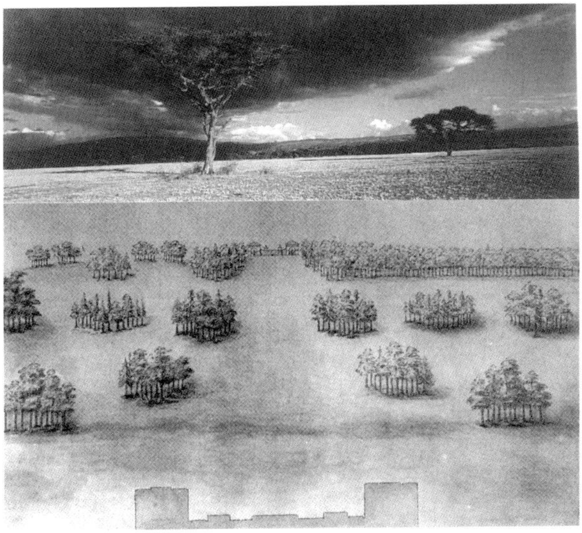

FIGURE 7 A typical savannah landscape. An eighteenth-century plan for the gardens of Holkam Hall in Norfolk displaying the characteristic savannah pattern of scattered tree shelter amidst open grassland (from Barrow [6]). (Holkam Hall, Norfolk, drawing (c. 1738) proposed planting of the north lawns by William Kent (Holkam Hall) Country Life Books (UK).)

and refuge. Many appealing images of tree houses, alcoves and inglenooks, *The Little House on the Prairie*, the castle in the mountains, and so forth, feed off this attraction [6]. Now go to the average art gallery or ask the average person in the street what sort of art works they like and you will find that landscapes that contain prospect and refuge symbols feature prominently. An example is shown in figure 8.

Of course, this reasoning need not apply to those who *study* art. They can overlay this default bias with learned appreciation and other experience. Becoming more speculative, a sensitivity for sunsets might be expected to pay off. Sunset indicates the imminent approach of twilight (when two sets of predators are around) and night. In general, a liking for "good" environments which are safe to enter and allow clear vision is likely to have greater survival value than an attraction for dark forests in which predators lurk unseen around corners [22, 23]. This approach teaches us why we do not like certain types of "concrete jungle" building projects. Huge concrete buildings which look as though they have landed from outer space with no obvious point of entry do not create a desire to enter, like the all too familiar example shown in figure 9.

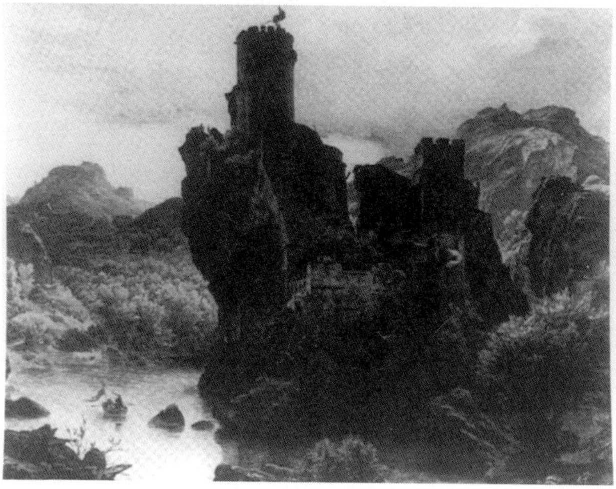

FIGURE 8 A landscape scene displaying prospect and refuge symbols, C.F. Lessing's *Castle on the Rocks*, 1828. (Reprinted by permission of Staatliche Museen zu Berlin—Preussischer Kulturbesitz, Nationalgalerie. Photo: Klaus Goeken/bpk.)

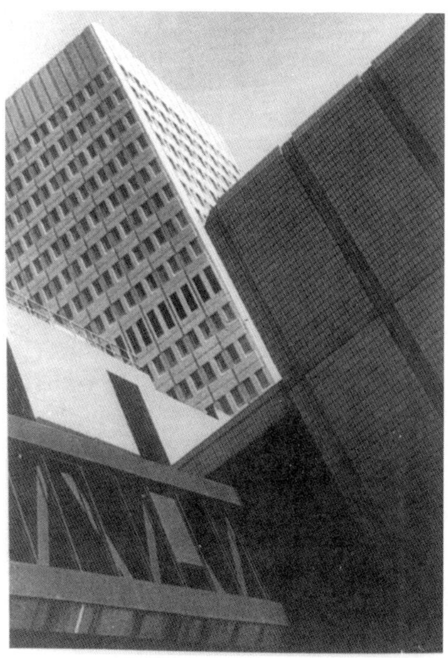

FIGURE 9 An uninviting building. (Reprinted by permission of J. Phillips (UK).)

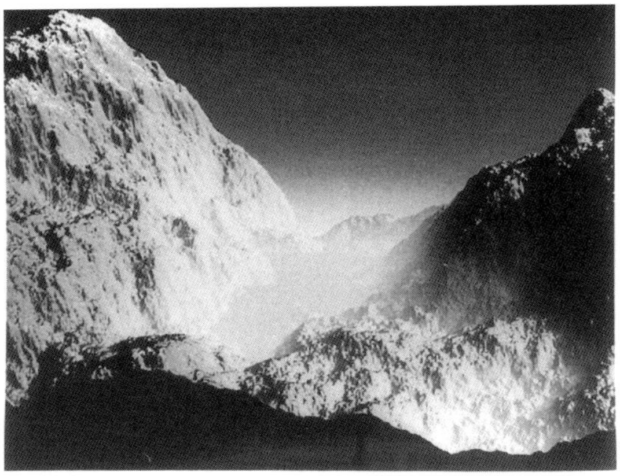

FIGURE 10 A fractal landscape generated by Richard Voss (from Barrow [6]). It displays the statistical variations with scale of genuine landscapes but lacks overt refuge symbols. (Reprinted by permission of R. F. Voss (USA).)

Likewise, if we are confronted with computer-generated landscapes [14][1] the prospect and refuge approach provides an important ingredient when it comes to understanding what it is we do and don't like about fractal landscapes, as can be appreciated by viewing figure 10. The addition of a small refuge symbol can transform that response. When we raise the question "Is computer art really art?" and set about evaluating examples we need to bear in mind this evolutionary bias that we possess. It is extremely influential.

It is also interesting to think more widely about the way in which art allows us to experience dangerous or unusual environments without risk. The combination of our imaginations with artificially-generated visual images is a powerful way of educating the brain and widening our experience in safe ways. It is striking how much we enjoy artificial forms of controlled danger: the ghost train, horror movies, roller coasters and computer games.

Similar considerations might apply to other common traditional artistic subjects. Still life studies of ripe fruit and flowers play on our unsurprising sensitivity to edible food and fruitfulness. The case of flowers is less obvious than that of the fruit though. We do not eat flowers. But flowers act as an important indicator of the identity of flora. Without flowers everything is chlorophyll green. A sensitivity to flowers will pay off compared to insensitivity. It will endow its owners with greater and more efficient discrimination in gathering.

As a last application of this reasoning we should of course make reference to our liking for symmetry—particularly lateral symmetry—the influence of which we saw in Birkhoff's attempts to quantify our measure of aesthetics. A ready appreciation of lateral symmetry is a good way of picking out living things from

[1]This fractal landscape was provided by Richard Voss, reproduced from Barrow [6] color plate 11.

non-living things in a crowded field of view. Living things display left-right symmetry externally (although not internally), but not up-down symmetry typically (because of gravity), and not front-back symmetry (if they move). A sensitivity for left-right symmetry should therefore be adaptive. It helps pick out potential mates, predators and prey. We see its legacy in the way that our superficial evaluation of human beauty is influenced by bodily and facial symmetry. Plastic surgeons are paid large sums of money to restore and enhance it.

4 THE CRITICAL APPROACH

One art form that is more amenable to analysis than most is music. At first one might have thought that the psychobiological approach would work well. However, unfortunately there does not seem to be a *single* obvious survival-enhancing ancient activity of which musical appreciation or music-making is an obvious by-product. The problem is there are lots of such activities. We could reasonably imagine music to be a by-product of imitating natural sounds (as camouflage), an activity to enhance group solidarity, timing and coordination skills, signaling, rhythmic drumming, inspiring warriors for battle, dance, mating calls (Charles Darwin's choice [10]), or the result of some other form of sexual selection (for further discussion and references, see Barrow [6]).

I want to suggest a different source that goes deeper beneath the appearances (sounds?) of music and suggest that there is evidence for something else from the common statistical character of the musics that different human cultures make and appreciate.

If we examine the spectrum of musical sounds that characterises musical traditions in a host of Western and non-Western musical traditions then it has been claimed that they all display a characteristic averaged power spectrum that is close to being self-similar and falls inversely as the first power of the sound frequency, f [20, 21]. This is the spectrum of so called $1/f$ ("one-over eff") noise, or "pink" noise, well known to sound engineers. An inverse-square fall-off, $1/f^2$, is characteristic of over-correlated Brown(ian) noise whilst no dependence, $1/f^0$, is characteristic of completely uncorrelated "white" noise. Pink noise displays a happy medium. It combines an optimal degree of predictability and unpredictability together with correlations on all time intervals. We clearly find this appealing to the ear and the brain. Completely random noise is unappealing (except to trained students of the genre) and encourages the pattern recognition programs in the brain to turn off (perhaps this is why "white noise" machines are such popular sleep inducers) whilst too much correlation (like doh, ray, me, far, so,...) is tedious. However, we know the $1/f$ spectrum, characteristic of natural flicker noise, may arise in any dynamical musical system at low frequencies. The studies of Voss and Clarke sample long-time correlations that average over different movements, pieces, style, and composers—especially in the samples from radio stations. Other commentators have argued that it is important to restrict the intervals to be no greater than single pieces of music [13, 17]. In the light of these concerns a more detailed study has been carried out of Western musical compositions by Boon and Decroly [9]. They investigate the power-spectrum of many musical compositions over the frequency interval 0.03Hz $< f <$ 3.0Hz and find that there is a good fit by a power-law $1/f^n$ but with $1.79 \leq n \leq 1.97$ rather than $n = 1$. This is very closer to the correlated Brownian noise ($n = 2$) spectrum rather than the pink noise seen over longer time intervals. However, it is clear that this power-spectrum analysis is sensitive only to a very particular statistical aspect of the musical dynamics. If we were to play the music backwards (or even turn it upside down!) the power spectrum would look the same. Other measures of the intrinsic complexity of the music need to be incorporated into these studies to capture more of the complexity that engages the mind.

The results of these studies then provide a basis for investigating whether there are any clear evolutionary advantages to heightened sensitivity to sound signals with these characteristic spectra. If so, then we could view one aspect of our particular musical appreciation as a by-product of adaptations that have developed for other reasons in the distant past.

The studies by Boon and Decroly are confined to Western music. It would be interesting to extend them to non-Western traditions. It would also be of interest to investigate the spectrum of whale song. There have been many speculations about its significance. The determination of its acoustic spectrum might shed some light on its function (if any).

In recent years the sand-pile paradigm of Per Bak [3] has been much used as a simple example for the creation and maintenance self-organised complexity. The critical slope of the sand pile is maintained by avalanches of sand that (ideally) occur on all scales from that of a single grain up to the size of the pile. As the sand pile grows from a few grains on a horizontal surface it does so by a collection of individually chaotic processes (individual sand grain trajectories) that have a more and more extensive effect on the pile as the critical slope is approached. At the critical slope the pile is most sensitive to perturbations and in this way it is possible to exercise influence over the whole pile and maintain a "self-organized" critical state. We might speculate that something 'critical' happens in the artistic experiences that we like. As a result, small nuances in performance, different interpretations of the music, different recordings, concert auditoria, or arrangements, all produce a noticeably different aesthetic experience. Interesting music is thus music that is worth listening to more than once. Its near criticality ensures that something new will be evident each time we hear it. Nor is this interpretation restricted to music. One criterion for sifting good art (books, paintings, sculpture,...) from mediocre is whether it is worth reading or viewing again. Shakespeare's plays have this bottomless quality despite constant performance and reading for hundreds of years. Perhaps we just like to keep our senses on the edge?

ACKNOWLEDGMENTS

I would like to thank the organisers for their invitation to participate in this meeting, the fellow participants for their stimulating talks and discussion, and the Director and staff of the Abisko research station for their unfailing hospitality.

REFERENCES

[1] Appleton, J. *The Experience of Landscape.* New York: Wiley, 1975.
[2] Appleton, J. *The Symbolism of Habitat.* Seattle, WA: University Washington Press, 1990.
[3] Bak, P. *How Nature Works.* Oxford University Press, Oxford, 1996
[4] Bak, P., C. Tang., and K. Wiesenfeld. "Self-Oorganized Criticality." *Phys. Rev. A* **38** (1988): 364.
[5] Barkow, J. H., L. Cosmides, and J. Tooby, eds. *The Adapted Mind.* New York: Oxford University Press, 1992.
[6] Barrow, J. D. *The Artful Universe.* Oxford: Oxford University Press, 1995.
[7] Berleyne, D. E. *Aesthetics and Psychobiology.* New York: Appleton-Century-Croft, 1971.
[8] Birkhoff, G. *Aesthetic Measure.* Cambridge, MA: Harvard University Press, 1933.
[9] Boon, J. P., and O. Decroly. "Dynamical Systems Theory for Music Dynamics." *Chaos* **5** (1995): 501.
[10] Darwin, C. *The Expression of the Emotions in Man and Animals.* London: Murray, 1872.
[11] Eigen, M., and R. Winkler. *Laws of the Game.* New York: Knopf, 1991.
[12] Hjort, B. "The Evolution and Ecological Significance of the Aesthetic Experience." *Man-Environment Systems* **9** (1979): 209.
[13] Klimontovich, Y. L., and J. P. Boon. "Natural Flicker Noise (1/f Noise) in Music." *Europhys. Lett.* **3** (1987): 395.
[14] Mandelbrot, B. *The Fractal Geometry of Nature.* San Francisco, CA: W. H. Freeman, 1977.
[15] Moles, A. *Information Theory and Aesthetic Perception.* Urbana, IL: University Urbana Press, 1966.
[16] McAllister, J. W. *Beauty and Revolution in Science.* Ithaca: Cornell University Press, 1996.
[17] Nettheim, N. "On the Spectral Analysis of Melody." *Interface—J. New. Mus. Res.* **21** (1992): 135.
[18] Orians, G. H. "Habitat Selection: General Theory and Applications to Human Behaviour." In *Evolution of Human Social Behaviour*, edited by J. Lockard, 49–66. New York: Elsevier, 1980.
[19] Orians, G. H. "An Ecological and Evolutionary Approach to Landscape Aesthetics." In *Landscape Meanings and Values*, edited by E. Penning-Rowsell and D. Lowenthal, 3–35. London: Allen and Unwin, 1986.
[20] Voss, R., and J. Clarke. "1/f Noise in Music and Speech." *Nature* **258** (1975): 317.
[21] Voss, R., and J. Clarke. "1/f Noise in Music: Music from 1/f Noise." *J. Acoust. Soc. America* **63** (1978): 258.
[22] Wilson, E. O. *Biophilia.* Cambridge, MA" Harvard University Press, 1984.
[23] Wilson, E. O. *Sociobiology: The New Synthesis.* Cambridge MA: Harvard University Press, 1975.
[24] Wilson, E. O. *Consilience.* Knopf, NY, 1998.

Art and Complexity
J. Casti and A. Karlqvist (editors)

Complexity and Aesthetics: Is Good Art "Complex" Art?

John L. Casti

1 WHAT IS ART?

In the theaters of both the West End of London and on Broadway, one of the big hits of the current theater-going season is a play called *Art.* The surface theme of the play revolves about the main character's purchase of a totally white-washed canvas for an outrageous sum. The character is then faced with trying to justify his extravgance to his friends by arguing the canvas's artistic merits.

Philsophically, *Art* is interesting for its exploration of the eternally vexing issue of just what is it *exactly* that distinguishes a piece of canvas and oil paint, like Jan Vermeer's *Allegory of Painting* as "art," while denying that label to something like the blank canvas in *Art?* Or what combination of words on the page lead to the claim that Dostoevsky's *Crime and Punishment* is a work of art, while no such claims are made for an alternate arrangement of words, such as John Grisham's *The Client?* It's instructive to see what some famous thinkers of the past have had to say about the matter.

Both Plato and Aristotle held to a representational theory of art, in which artworks imitate real physical objects. But they differed radically on the matter of whether it was possible to gain either intellectual or practical knowledge about real-world things from their art-world representations.

In Plato's philosophy, true reality resides in the Eternal Forms or Ideas that make it possible to understand ordinary physical objects. Thus, according to Plato the way to gain knowledge is to directly encounter these "platonic" Forms. But since artworks are only imitations of physical objects, which are themselves only derivatives of the Forms, a work of art cannot provide us with any knowledge. As Plato described it in Book 10 of *The Republic:* "They [artworks] are at the third remove" from reality. So, for Plato art was not a source of knowledge or even of reliable opinion about objects of the real world.

Aristotle, although sharing Plato's view of art as presenting likenesses of things, argued that it is natural and beneficial for humans to learn by imitating and from carefully crafted imitations. In the *Poetics,* Aristotle noted that tragic poetry, unlike history, often expresses general truths, not just the facts of what actually took place. Rather, poetry tries to convey a feeling for what is likely to happen, generalizable truths about the sorts of things that probably or necessarily occur. But, he went on to say, it may be difficult to understand events that match these general truths, especially when these events are taking place in real time all around us.

As a result, Aristotle suggests that by composing an imitation of an action that is carried out on stage, the dramatist can display the same truth that is being shown by the real action, but in circumstances helpful to learning about the situation. The real action, possibly containing real death, tragedy and destruction, might distract us from the chance to learn. But a suitably idealized imitation of the action may allow us to comprehend the principles that govern human activity. This is somewhat analogous to studying, say, the human heart or kidney. A plastic laboratory model of these organs might facilitate learning about their typical structure more effectively than dissecting the real heart or kidney of an anonymous corpse. In this case, the model reproduces and emphasizes the organ's essential structure and general features, but it eliminates the peculiarities and possibly repulsive and distracting aspects of a real organ.

Both Plato and Aristotle recognized that an essential aspect of art is that it is *different* from real things. Their views part company only on the point of whether we can learn about real things from this difference. For example, Plato would argue that there is nothing to be learned about late-nineteenth-century Parisian life from gazing upon Renoir's famous painting *Luncheon of the Boating Party.* Aristotle, on the other hand, may well argue that this painting encapsulates an enormous amount of information about how people of a certain social class interacted and how they lived in *fin-de-siècle* Paris. Nevertheless, the portrait of Parisian life shown by Renoir is certainly not the real thing, and to believe it is would be like having a member of the audience jump up and call for the police during the scene in Shakespeare's play *Othello* when Othello strangles Desdemona on stage. In both the play and the painting, a crucial aspect of understanding the artwork lies in realizing that art objects must be different from real things.

Interestingly enough, postmodern artists try to reduce the distance between art and real things. As an illustration, consider the artist Robert Indiana who paints pictures of bull's-eye targets that are at the same time real targets and imitations of real targets. Now suppose you hung a real target next to such a painting. Would it be acceptable for an archer to shoot arrows at the Indiana painting? Or would an art afcionado object that you should restrict your shooting only to the target, even though the target and the painting look exactly alike? Does the Robert Indiana painting tell us anything about real targets by imitating them in paint on a canvas? That is, do we learn anything about the real system from a model that is indistinguishable from it? Hard questions.

What about complexity? Is there are consistent relationship between the perceived "quality" of a piece of art and any reasonable measure of its complexity? Is a complex artwork more aesthetically satisfying than one that is "simple"? To even pose this question implies that we have some type of complexity measure that is intrinsic to the piece of art itself, and which doesn't depend on the person observing the painting, sculpture or whatever other type of artwork may be under consideration. Dubious as such an hypothesis may be, let's follow through its implications.

2 GENERATIVE ART

One of the most well-developed approaches to characterizing the complexity of patterns is algorithmic complexity theory, which takes the complexity of a pattern to be the length of the shortest computer program that will reproduce the pattern under study. Since this shortest length may vary, depending on the particular computing machine and computer language used to describe the situation, it's customary to fix these variables by assuming that we use a universal Turing machine, together with the limited set of standard instructions appropriate for such a gadget. I won't worry about these technical fine points here, as they're not important for the issues we want to explore. In fact, instead of a universal Turing machine and its langauge, let's consider a real computing machine and the popular programming language LISP to look at the complexity of some pieces of art.

A few years ago, computer scientist Karl Sims had the idea of regarding expressions in the LISP programming language as genotypes in an evolutionary process [5]. When executed on the computer, the result can then be thought of as the phenotype generated by the LISP expression. Sims's goal was to create a process of artificial evolution using these symbolic expressions. In this process, the LISP expressions open up the opportunity for the emergence of a genuinely new developmental rule or parameter value beyond the boundaries of what may have been set by the programmer at the outset of the experiment. It's not really important for us to know the exact meaning of these LISP expressions, other than that each of them takes a specific number of arguments and returns an image of black-and-white or color values for each pixel on the computer's terminal screen. Nevertheless, it is of interest to examine a few of the expressions Sims used just to get a feel for what he had in mind with these experiments.

In Sims's work, the LISP expressions could be formed of combinations of any of the following common LISP functions:

1. *X*
2. *Y*
3. *(abs X)*
4. *(mod X (abs Y))*
5. *(and XY)*
6. *(bw-noise .2 2)*
7. *(color-noise .1 2)*

8. *(grad-direction(bw-noise .15 2).0 .0)*
9. *(warped-color-noise(*X .2) Y .1 2)*

Color plate 1 shows the type of image each of these functions produces from an initially black square, where the functions 1–9 are read left to right, top to bottom.

Sims began his experiments by creating a LISP expression that combined a random number of these functions. Such an expression was then translated by a LISP interpreter into a graphic image, the phenotype associated with this symbolic genotype. Since LISP expressions can be written as tree structures, the mutation of such an expression proceeds by traversing the tree, node-by-node, and applying one or another mutation schemes at each node. A typical such scheme might say that if the node is a function like *(abs X),* it might mutate into a different function like, for instance, *(cos Y).* In addition, symbolic expressions can be reproduced with "sexual combination" by combining the parent expressions in various ways. Color plate 2 shows the effect of 19 mutations of this sort on a parent in the upper left-hand corner. This plate shows only the surface image of a three-dimensional structure that Sims created by adding a volume texture operation that calculates color values for each point in three-dimensional space.

By starting with randomly generated genomes and applying a variety of types of mutations, Sims played the role of Father Nature, selecting those mutations that would be allowed to live on to the next generation. After anywhere from 5 to 20 generations, a remarkable set of graphic images emerged. Color plate 3(a)–(c) shows a small sample of Sims's art gallery, along with the LISP genotypes that coded for these pictures. If you're wondering why there is no genotype displayed for part (c) of the plate, it is because this phenotype was created before Sims added a genotype-saving subroutine to his program. Part (c) is what one might call an extinct species in this world of evolutionary art forms. It's a point worth pondering to note here how 186 characters of the alphabet can code for a complicated artistic object like part (b) of the plate.

So what does this exercise tell us about complexity and art? Is the picture in part (b) of plate 3 more aesthetically satisfying than that in part (a), for instance? It's certainly more complex, at least by the measure of program length. But perhaps program length isn't such a good measure after all, since by such a criterion for complexity the most complex objects are those that have no pattern at all! The totally random objects have the highest complexity, and completely randomness is definitely not what we have in mind when it comes to separating great works of art from the pretenders. Perhaps another standard of complexity is called for.

3 CONNECTIVITY AND ART

An almost self-evident feature of works of art of all types is that they represent a connective structure at many different hierarchical levels. The colors and shapes of a painting are integrated into substructures, which in turn form parts of even larger substructures until we encompass the entire work in one, grand pattern. To

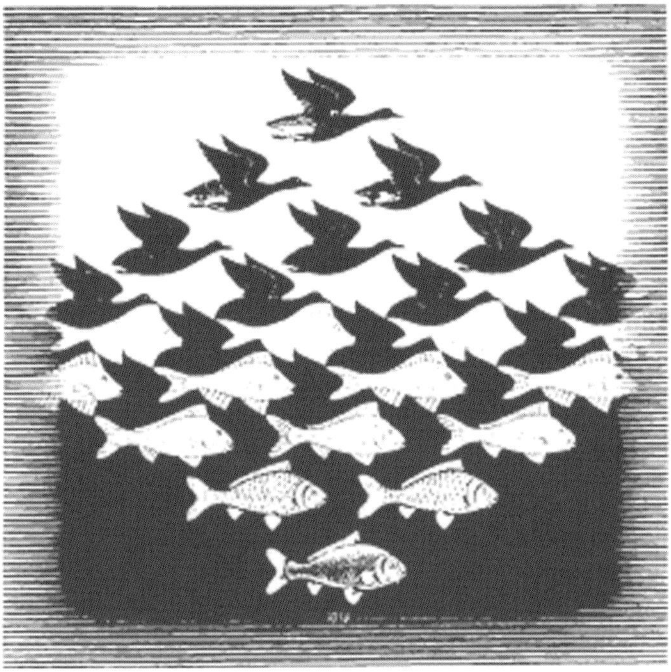

FIGURE 1 M. C. Escher, *Sky and Water* (1938). (M. C. Escher's "Sky and Water I"

a mathematician, this kind of pasting together of local patterns to form global structures is very reminiscent of what topologists do in order to create global geometries like that of a planet, or even the universe, from individual patches of more-or-less flat geometries of the type Euclid so admired. Some of the same techniques can be employed to study the structure of artworks, as well.

Nowadays it's almost impossible to walk into the office of a scientist or mathematician without seeing an engraving or two by the well-known Dutch artist, M. C. Escher (1898–1971) hanging on the wall. Escher is noted for the remarkable geometrical precision of his work, as well as for its deep connections with mathematical concepts, especially those in group theory. Here we examine one of his more famous works using a collection of tools borrowed from algebraic topology based on the idea of a simplicial complex. This is an object composed of points, lines, triangles and so forth, which forms the mathematical skeleton for constructing the algebraic structure of continuous spaces of all types.

A good illustration of the use of simplicial complexes to capture abstract structure is provided by Escher's famous engraving *Sky and Water,* shown in figure 1. Here we see a collection of what appear to be geese gradually being transformed into fish as the picture is scanned continuously from top to bottom.

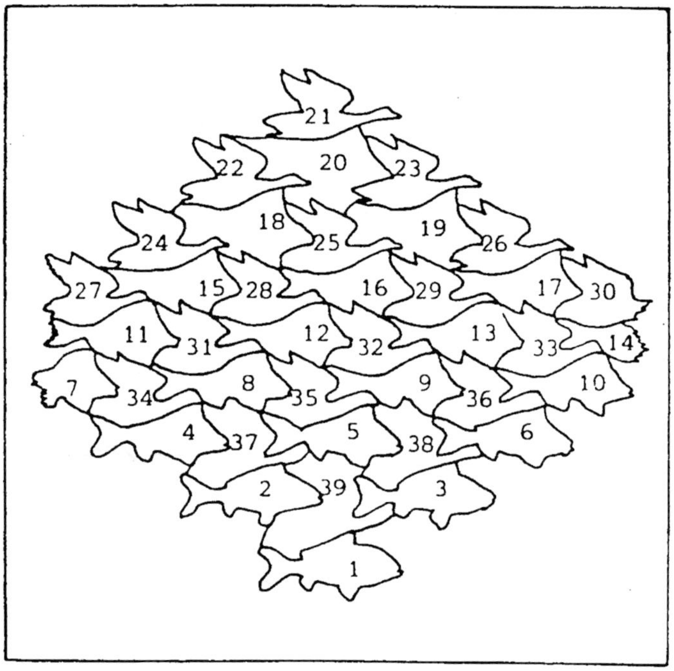

FIGURE 2 Shapes in *Sky and Water.*

At the same time, we also see a smooth transition from figure to ground as the shapes constituting the geese become background for the swimming fish. Our goal is to capture some of the structure of these transitions using topological quantities.

The first step is to identify relevant sets X and Y whose elements are somehow "related" to each other through the engraving. The relation, whatever it may be, somehow encapsulates an important part of the connective structure of *Sky and Water.* It doesn't take much reflection to see that the picture is really a statement about the relationship between various geometrical shapes (the birds, fishm and their intermediate forms), and features that pertain to the identification of the shapes as being birdlike, fishlike or something in between. Figure 2 identifies 39 different shapes that appear in the picture. So we let the elements of the set $Y = \{y_1, y_2, \ldots, y_{39}\}$ be these shapes.

As for the set X, its elements are the following collection of 12 features, each of which plays a prominent role in the picture:

$$X = \{x_1, x_2, \ldots, x_{12}\},$$
$$= \{\text{scales, mouth, gills, fish-tail, fins, fish shape, eye, duck shape, two wings, feathers, beak, legs}\}.$$

We make the obvious choice for the relation λ linking the elements of X to those of Y. We define a pair (y_i, x_j) to be in the relation λ if and only if shape y_i displays feature x_j. It's algebraically convenient to represent this relation by an incidence matrix Λ, whose elements Λ_{ij} are either 1 or 0, depending on whether the pair (y_i, x_j) is in the relation λ or not in the relation, respectively. A suitable incidence matrix is given below:

λ	x_1	x_2	x_3	x_4	x_5	x_6	x_7	x_8	x_9	x_{10}	x_{11}	x_{12}
y_1	1	1	1	1	1	1	1	0	0	0	0	0
y_2	1	1	1	1	1	1	1	0	0	0	0	0
y_3	1	1	1	1	1	1	1	0	0	0	0	0
y_4	1	1	1	1	1	1	1	0	0	0	0	0
y_5	1	1	1	1	1	1	1	0	0	0	0	0
y_6	1	1	1	1	1	1	1	0	0	0	0	0
y_8	0	1	1	1	1	1	1	0	0	0	0	0
y_9	0	1	1	1	1	1	1	0	0	0	0	0
y_{10}	0	1	1	1	1	1	1	0	0	0	0	0
y_{11}	0	0	0	1	1	1	1	0	0	0	0	0
y_{12}	0	0	0	1	1	1	1	0	0	0	0	0
y_{13}	0	0	0	1	1	1	1	0	0	0	0	0
y_7	0	1	1	0	1	0	1	0	0	0	0	0
y_{21}	0	0	0	0	0	0	1	1	1	1	1	1
y_{22}	0	0	0	0	0	0	1	1	1	1	1	1
y_{23}	0	0	0	0	0	0	1	1	1	1	1	1
y_{24}	0	0	0	0	0	0	1	1	1	1	1	1
y_{25}	0	0	0	0	0	0	1	1	1	1	1	1
y_{26}	0	0	0	0	0	0	1	1	1	1	1	1
y_{28}	0	0	0	0	0	0	1	1	1	1	1	1
y_{29}	0	0	0	0	0	0	1	1	1	1	1	1
y_{31}	0	0	0	0	0	0	1	1	1	0	0	0
y_{32}	0	0	0	0	0	0	1	1	1	0	0	0
y_{33}	0	0	0	0	0	0	1	1	1	0	0	0
y_{27}	0	0	0	0	0	0	1	1	1	1	1	0
y_{30}	0	0	0	0	0	0	0	0	1	1	0	0
y_{34}	0	0	0	0	0	0	0	1	0	0	0	0
y_{35}	0	0	0	0	0	0	0	1	0	0	0	0
y_{36}	0	0	0	0	0	0	0	1	0	0	0	0
y_{37}	0	0	0	0	0	0	0	1	0	0	0	0
y_{38}	0	0	0	0	0	0	0	1	0	0	0	0
y_{39}	0	0	0	0	0	0	0	0	0	0	0	0
y_{14}	0	0	0	1	0	0	0	0	0	0	0	0
y_{15}	0	0	0	0	0	1	0	0	0	0	0	0
y_{16}	0	0	0	0	0	1	0	0	0	0	0	0
y_{17}	0	0	0	0	0	1	0	0	0	0	0	0
y_{18}	0	0	0	0	0	0	0	1	0	0	0	0
y_{19}	0	0	0	0	0	0	0	1	0	0	0	0
y_{20}	0	0	0	0	0	0	0	0	0	0	0	0

Using Λ, we and obtain the following connective structure in *Sky and Water:*

$$\begin{array}{lll}
q = 6: & Q_6 = 1, & \{y_1 - y_6\}, \\
q = 5: & Q_5 = 2, & \{y_1 - y_6, y_8 - y_{10}\}, \{y_{21} - y_{26}, y_{28}, y_{29}\}, \\
q = 4: & Q_4 = 2, & \{y_1 - y_6, y_8 - y_{10}\}, \{y_{21} - y_{29}\}, \\
q = 3: & Q_3 = 2, & \{y_1 - y_{13}\}, \{y_{21} - y_{29}\}, \\
q = 2: & Q_2 = 2, & \{y_1 - y_{13}\}, \{y_{21} - y_{29}, y_{31} - y_{33}\}, \\
q = 1: & Q_1 = 2, & \{y_1 - y_{13}\}, \{y_{21} - y_{33}\}, \\
q = 0: & Q_0 = 1, & \{\text{all}\}.
\end{array}$$

What this means is that at dimension level 6, there is but a single component in the relation. This component consists of the six shapes $y_1 - y_6$, which are all connected to each other by sharing seven or more features from the set X. (Note: Since one point constitutes an object of dimension 0, we need $k + 1$ points to determine a k-dimensional object.) At dimension level 5, the engraving splits into two disjoint components, the shapes $y_1 - y_6, y_8 - y_{10}$ and the shapes $y_{21} - y_{26}, y_{28} - y_{29}$. This is an indication that if your mind is capable of appreciating objects of dimension 6 or higher, you see the engraving as simply the single component at that level. But if you can only "see" in dimension 5 or below, *Sky and Water* splits into two disconnected components. Finally, at dimension level 0 there is the single components consisting of all the shapes. This means only that when we look at the engraving, we see it as a single, unified piece of work and not two (or more) disconnected components.

The foregoing analysis of the engraving focuses attention upon the shapes, showing that the principal shapes in the picture are the "fish" shapes $y_1 - y_6$, followed by the "bird" shapes $y_{21} - y_{26}$. This is a fairly obvious conclusion; nevertheless, it's satisfying to reach it via our systematic procedures. At the intermediate levels of connectivity $1 \leq q \leq 5$, we see that *Sky and Water* breaks down into two disconnected pieces, essentially fishlike and birdlike shapes, whereas at the extreme levels $q = 6$ and $q = 0$ we have a fully integrated picture.

Those readers familiar with other works by Escher will recognize that *Sky and Water* is typical of many of his engravings, which feature a smooth passage from one type of figure to another accompanied by a transition from figure to ground. The techniques introduced above, together with the deeper and more refined methods presented in literature cited in Atkin [1], Casti [3], and Johnson [4], offer the basis for a systematic analysis of many aspects of Escher's style and form.

4 IN THE EYE OF THE BEHOLDER

The use of both algorithmic complexity and hierarchical analysis has shed some light on the matter of complexity and aesthetics, if only to suggest that more complex is not necessarily better. But it's easy to see from this discussion that there's a lot more to aesthetics than just complexity—especially any notion of complexity that rests on the assumption that works of art have an intrinsic complexity, independent of the perceptions and prejudices of an external observer. After all is said and done, a purely syntactic notion of intrinsic complexity is likely to end up telling us nothing more than that art, like beauty of every type, resides much more in the eye of the beholder than in the beholden.

REFERENCES

[1] Atkin, R. *Mathematical Structure in Human Affairs.* London: Heinemann, 1974.
[2] Casti, J. *Would-Be Worlds.* New York: John Wiley & Sons, 1997.
[3] Casti, J. *Reality Rules*, vol. II. New York: John Wiley & Sons, 1992.
[4] Johnson, J. *Combinatorial Structure in Digital Pictures,* NERC Remote Sensing Project, Discussion Paper No. 1, Center for Configurational Studies, The Open University, Milton Keynes, UK, January 1985.
[5] Sims, K. "Artificial Evolution for Computer Graphics." *Computer Graphics* **25**(4) (1991): 319–328.

Art and Complexity
J. Casti and A. Karlqvist (editors)

What Lies Between Order and Chaos?

James P. Crutchfield

What is a pattern? How do we come to recognize patterns that we've never seen before? Formalizing and quantifying the notion of pattern and the process of pattern discovery go right to the heart of scientific practice. Over the last several decades science's view of nature's lack of structure—its unpredictability—underwent a major renovation with the discovery of deterministic chaos. Behind the veil of apparent randomness, many processes are highly ordered, following simple rules. As the new millennium begins, tools adapted from the theory of computation will bring empirical science to the brink of automatically discovering patterns and quantifying their structural complexity. For example, rather than interpreting a data stream according to a given model, we look at a model stream. The regularities found in the *way* models improve with learning is the basis for inferring universal laws on how complexity arises from the interaction of order and chaos.

[A popular essay solicited to appear in *The Sciences*, New York Academy of Sciences, New York (1994). Sadly, NYAS no longer publishes *The Sciences*. February 2002: This version is somewhat updated from the original, written in 1992: citations have been added and dated comments edited to read less obviously a decade old.]

1 INTRODUCTION

During the Summer of 1927 Balthasar van der Pol, a Dutch engineer, listened to the tones produced by a neon glow lamp coupled to an oscillating electrical circuit.

Lacking modern electronic test equipment, he monitored the circuit's behavior by listening through a telephone ear piece. In what is probably one of the earlier experiments on electronic music, he discovered that, by tuning the circuit as if it were a musical instrument, fractions or subharmonics of a fundamental tone could be produced [27]. This is markedly unlike common musical instruments—such as the flute which is known for its purity of harmonics, or multiples of a fundamental tone. As van der Pol and a colleague reported in the September 10th issue of the British journal *Nature* that year "the turning of the condenser in the region of the third to the sixth subharmonic strongly reminds one of the tunes of a bag pipe."

There is a curious aside in the report, however. The experimenters noted that when tuning the circuit "often an irregular noise is heard in the telephone receivers before the frequency jumps to the next lower value." We now know that van der Pol had listened to deterministic chaos: the noise was produced in an entirely lawful, ordered way by the circuit itself. The *Nature* report stands as one of its first experimental discoveries. Other concerns were on the experimenters minds, for the report immediately continues "However, this is a subsidiary phenomenon...." With this remark their primary interest in the design of stable radio oscillators led them away from discovering the order in the chaos.

Much of our appreciation of nature depends on whether our minds or, more typically these days, our computers are prepared to discern its intricacies. When confronted by a phenomenon for which we are ill-prepared, we often simply fail to see it, though we may be looking directly at it.

Indeed, what is a "pattern" in nature? More to the point, how do we come to notice a "pattern" we've never seen before? How can we ever see past our own assumptions? Formalizing and quantifying the notion of pattern goes right to the heart of scientific practice. Over the last several decades our view of nature's lack of structure—its unpredictability—underwent a major renovation with the discovery of deterministic chaos. As the new millennium begins, ideas adapted from the theory of computation will bring empirical science to the brink of automatically discovering patterns and quantifying their structural complexity. One guide to this will be universal laws on how complexity arises from the interaction of order and chaos.

2 BACKGROUND

Van der Pol and his colleague J. van der Mark apparently were unaware that the deterministic mechanisms underlying the noises they'd heard had been rather keenly analyzed three decades earlier by the French mathematician Jules Henri Poincaré in his efforts to establish the orderliness of planetary motion. The motion of the planets about the sun is one of the hallmarks of regularity and predictability. But when mathematicians and physicists attempted to finally prove this, trouble arose. At the very close of the nineteenth century Poincaré in his treatise *Nouvelles Méthodes des Mécanique Céleste* focused on the collective motion of the sun, a planet, and a moon—the famous "three-body problem" [23]. After nearly 1500 pages of detailed successful analysis and simplification, he ran into deep complications in solving for their motions:

> If one seeks to visualize the pattern formed by these two [solution] curves...[their] intersections form a kind of lattice-work, a weave, a chain-link network of infinitely fine mesh; ... One will be struck by the complexity of this figure, which I am not even attempting to draw. Nothing can give us a better idea of the intricacy of the three-body problem, and of all the problems of dynamics in general....

It is impossible to know his state of mind when, at the end of a Herculean effort to establish mathematically the observed fact of the solar system's stability, Poincaré realized the daunting complexity of the task. He dryly notes that it was "a point which gave me a great deal of trouble."

The puzzle of deterministic chaos is just one example from twentieth-century science that shows how the limitations of human understanding make nature appear "noisy," complicated, and unpredictable. One immediately thinks of quantum mechanics, another legacy from the early part of the twentieth century, as putting severe limits on purely objective measurements of nature [15]. But even in the rarefied world of the foundations of mathematics similar roadblocks appeared. Kurt Gödel demonstrated that logical consistency had to be traded-off against one's ability to prove the possible "truths" within a formal system—even a system as simple as arithmetic [21]. Alan Turing then showed, more concretely, that well-formulated questions in a formal system may have no constructive answers [26]. More recently, Gregory Chaitin has argued that there is an irreducible element of randomness in mathematics that limits its effectiveness [3].

Psychology and philosophy in the twentieth century were punctuated by a series of similar disappointments. Limitations and the complication they engender permeate much more than just mathematics and physics. Freud, to take one example, called into question the Western concept of a whole and knowable self controlling the mind. He viewed all apparently spontaneous arbitrary actions as being at the beck and call of the unconscious, of which one could have no knowledge [12, 13]. Derrida then deconstructed the remaining notion of a self, which was based, he thought, on erroneous notion of a metaphysical presence. Derrida, it seems, would have us believe that there is chaos in our own houses [11].

These limitations suggest that humans are strongly predisposed to make many unjustified, often unspoken, simplifying assumptions about nature and experience. At first, these assumptions are frustrated and the world appears complicated, structureless, and random. Once they are finally acknowledged and become an object of study, a "new" limitation on our knowledge is discovered.

This list of limitations, which could be easily extended, paints a rather pessimistic picture of the progress of human knowledge. But it also raises a constructive question: If things are so complicated, how do we ever discover patterns and regularity? Is a hurricane's path really unpredictable or is there some hidden order that we do not yet appreciate? How can the lawfulness producing deterministic chaos ever be extracted, if its outward appearance is random?

These questions highlight the very activity by which scientists penetrate the veil of complication and distill new laws from experiments. How do scientists balance the need for order against nature's seeming chaos? As certainly as we have come to appreciate our limitations, this century has fostered an unparalleled increase in our knowledge of nature. Somewhat ironically, the realization of each limit, rather than being only a disappointment, often showed nature to be much richer than before, when seen through the dark sunglasses of simplifying assumptions. The rapid increase in knowledge suggests there may be some driving force behind this progress. Can we be scientific about the practice of science? Is there a dynamic of discovery?

Certainly, whatever this dynamic is, it is not unfamiliar to us. The structural anthropologist Claude Levi-Strauss describes the process as he experienced it during his first treks in the 1930s into the Amazon [18]:

> Seen from the outside, the Amazonian forest seems like a mass of congealed bubbles, a vertical accumulation of green swellings; it is as if some pathological disorder had attacked the riverscape over its whole extent. But once you break through the surface-skin and go inside, everything changes: seen from within, the chaotic mass becomes a monumental universe. The forest ceases to be a terrestrial distemper; it could be taken for a new planetary world, as rich as our world, and replacing it.
>
> As soon as the eye becomes accustomed to recognizing the forest's various closely adjacent planes, and the mind has overcome its first impression of being overwhelmed, a complex system can be perceived.

Despite confronting what initially appears to be structurelessness, we seem to be able eventually to discover the hidden order.

One of the most fascinating spontaneous pattern discovery and learning processes is a child's acquisition of language. Imagine the trade-offs that an infant faces in balancing the initial apparent structurelessness of what it hears and its need to find order. Allison Gopnik, a child psychologist at UC Berkeley, has suggested that infants in their developmental succession of world views are like scientists, forming and testing hypotheses and rejecting those that are unhelpful, inconsistent, or too complicated [14].

Natural language itself shows a balance between order and randomness [5]. On the one hand, there is a need for static structures, such as a vocabulary and a grammar, so that two people can communicate. Without a prior agreement on these there is no basis for understanding; each and every utterance would be unintelligible to the listener—a common experience for the world traveler. On the other hand, there would be no need to communicate if spoken utterances were completely predictable by the listener. In this case the language would be a rigidly fixed structure with all possible sentences uniquely identified and identifiable. But humans use language (typically) to communicate new information—facts, ideas, feelings, and other states of mind. And so, there must be an unknown or unexpected element in communication as far as the listener is concerned, if they are to stay engaged. Then again the "new" element cannot be so dominant that the result is a jumble of phonemes, words, and sentences. Natural language as a changeable and dynamic system must be a balance of new information unpredictable by the listener and of order so that communication is understandable.

Is there a general principle that guides the dynamic balance of order and chaos? And what is the result of this balance? In his *Process and Reality* [28], the

British philosopher Alfred North Whitehead comments on the interplay of order and chaos in art:

> The same principle is exhibited by the tedium arising from the unrelieved dominance of fashion in art. Europe, having covered itself with treasures of Gothic architecture, entered upon generations of satiation. These jaded epochs seem to have lost all sense of that particular form of loveliness. It seems as though the last delicacies of feeling require some element of novelty to relieve their massive inheritance from bygone system. Order is not sufficient. What is required, is something much more complex. It is order entering upon novelty; so that the massiveness of order does not degenerate into mere repetition; and so that the novelty is always reflected upon a background of system.

So is it complexity that is the result of the balance of order and chaos? But what is complexity? Are there any general principles that govern the interplay of order and chaos, that aid in detecting structure and pattern? How does genuinely new information arise from a structureless universe? Finally, why do humans presume that there is order to be found in a chaotic, uncharted nature? Recent work has begun to elucidate this drive toward finding regularity in nature and, in particular, the trade-offs between order and chaos that occur in the process of acquiring new knowledge.

3 COMPLEXITY

The weather is often considered a prime example of unpredictable behavior. The simple truth, though, is that it is quite predictable. Over the period of one minute (say), one can surely predict it. With a glance out the nearest window to note the sky's disposition, one can immediately report back a forecast. To predict over one hour, one would search to the horizon, noting much more of the sky's prevailing condition. Only then, and not without pause to consider how that might change during the hour, would one offer up a tentative prediction. If asked to forecast two weeks in advance one would probably not even attempt the task. Why even look out the window? The necessary amount of information and the time to assimilate it for a two-week forecast would be overwhelming. Despite the long-term unpredictability, a meteorologist can write down the equations of motion for the forces controlling the weather dynamics in each case. In this sense, the weather's behavior is symbolically specified in its entirety. How does unpredictability arise in such a situation?

One meteorologist, Edward Lorenz of MIT, did analyze the equations governing weather dynamics with this question in mind [19]. Focusing on particularly simple deterministic equations, in 1963 Lorenz proposed a mechanism—the butterfly effect—that actively amplifies even the most microscopic, and uncontrollable, events to macroscopic proportions; this is *the* mechanism underlying *deterministic chaos.* Imagine that a meteorologist is allowed to use as much historical weather data and as much computer time as needed for a moderately accurate four-day forecast. What Lorenz found was that if the meteorologist tries to extend the forecast for one additional day, while maintaining the same degree of accuracy, *twice* as much historical data and computer time are required. The result is that, if deterministic chaos is present, there is an irreducible error in long-term predictions, since the required resources grow so quickly: If the resources required for predicting double with each additional day, extending the forecast by only ten days requires a thousand-fold increase, rapidly overwhelming any effort to forecast.

We now measure this degree of long-term unpredictability using the *entropy rate*, a quantity introduced by Claude Shannon in his theory of communication that measures the degree of surprise when receiving messages produced by some source [25]. The Russian mathematician Andrei Kolmogorov adapted this to view a deterministic chaotic system—such as the three-body problem—as an information source [16]: As observers, we are surprised when our predictions of its behavior fail. If we measure the state of a system to an accuracy of one part in a thousand and if the system doubles that measurement uncertainty every second, then after one second we know the system state to only one part in five hundred. In terms of Shannon's entropy rate, the system has produced one bit of information, since we can resolve only half as many distinct states. At that rate of information production, the system is completely unpredictable after only ten seconds.

Lorenz's work suggested that unpredictability was inherent in very large systems, such as the weather, not only in systems with a few components, such as the three-body problem analyzed by Poincaré. Their work left open the question of how a chaotic system is structured to support a given degree of unpredictability. In 1982 Norman Packard and I proposed that the structural complexity of a process, such as the weather, could be measured by the decay in one's ability to predict its behavior as one accumulates additional information [8]. We called this complexity the *excess entropy*, since it captured the initial apparent disorder above the long-term unpredictability.

To see how this works we envisioned a meteorologist making a succession of observations. Initially, before any measurements are made, the weather could be anything; the meteorologist is ignorant of the prevailing conditions and forecasts have nothing to do with the actual weather. It is highly unpredictable; the entropy rate apparent to the meteorologist is very large. After a few observations, though, the meteorologist knows the current condition and has the possibility of noticing regularities: Are the conditions changing? By how much and in what way? The additional information allows much better forecasts, certainly, than before observations were begun. As more information is accumulated through succeeding observations, the accuracy of forecasts continues to improve until the ceiling imposed by the weather's inherent unpredictability is reached.

The excess entropy was invented to monitor just how this increase to optimal forecasting comes about. To see how it differs from the entropy rate, which sets the ceiling on long-term unpredictability, consider three different types of weather. The first is a sunny day, with clear and calm skies. This weather behavior is very easy to predict: once we know the current wind velocity, temperature, and humidity, we forecast that they will continue. If we make further observations, there are no surprises; the entropy rate is zero, the system is not chaotic. We also come to notice the regularity very quickly. Just one observation of the temperature, wind velocity, and so on, is all that is required to set up the forecast. In this case, the excess entropy is low, since only a few observations are required to know the prevailing (exactly predictable) conditions.

The second example comes from the other extreme. Imagine we are in the crush of a horrendous storm, wild winds and sudden downpours pelt the land, with no chance of letting up. This weather behavior—the change in wind direction, the variation in local temperature and humidity—is very difficult to predict. We are maximally uncertain about the weather: we keep looking out the window for an update and are constantly surprised; the entropy rate is high. We come to appreciate this high unpredictability very quickly, after only a few observations. We also immediately realize that it's not really worth the effort to accumulate detailed observations and have our computers (say) develop a forecast, since the conditions are so changeable. In this case, as for the calm weather, the excess entropy is low. Independent of the weather's predictability, only a few observations are required to learn its condition (highly unpredictable). In other words, highly predictable *and* highly unpredictable behaviors are *simple*, since the method of forecasting is so straightforward. For the calm weather we simply report that our first observations will continue. For the stormy whether, we make our forecasts by flipping a coin. In both cases, after a while we don't even bother to look out the window.

The genuinely interesting cases fall between these two extremes. Instead of our forecasts being either exactly right or almost always wrong, imagine weather that regularly alternates between clear skies and cloud bursts. When it is clear, we certainly want to know this, since for that period our forecasts will be correct. It is also useful to know when the weather switches to being stormy. Since our forecasts, then, will be wrong on average, we can reduce our effort to predict and go back to simply guessing. To make optimal forecasts in this situation, we must monitor the weather closely: Is it clear or stormy? Since half our forecasts are wrong, the entropy rate is somewhere between zero and the maximal value: there are some elements that are predictable. But it takes a long time to appreciate just what those elements are and the amount of effort used to take advantage of them for optimal forecasts is quite high. The result is that the excess entropy is large, unlike that found at the extremes of predictability. This intermediate behavior is more "complex" than either extreme. One needs more observations to know the prevailing conditions, our models need to be more sophisticated, and the effort to forecast is larger. In short, more information is required for optimal prediction in this intermediate case.

These examples serve to illustrate a general principle that as one moves across the spectrum of predictability—from ordered to random behavior—the "complexity" is maximized in the middle. The excess entropy is one measure of how processes are structured and it is a necessary tool for our understanding how nature comes to appear more or less predictable to an observer. Since it was introduced, a number of similar proposals to measure "physical complexity" have appeared [29]. Like the excess entropy, each alternative attempts to capture the amount of information processing that a system employs to produce its unpredictability. Their main failing, however, is that they do not tell us *how* that information is processed.

As a first step to address this, in 1987 Bruce McNamara and I showed how one could extract from experimental data the underlying equations of motion, a compact representation of the governing forces [7]. Although our approach addressed certain issues of automated modeling, its main problem was that there appeared to be no way estimate from the symbolic equations of motion how much information processing was being performed by the system.

To remedy this, in 1989 Karl Young and I introduced a method to reconstruct from observations the hidden computational mechanisms underlying unpredictable behavior [9]. We adapted several ideas from the earliest days of computers, in particular those introduced by Noam Chomsky, the MIT linguist [4]. To Chomsky, the activity of building a grammar for a language was analogous to the construction of a scientific theory from experimental data. He proposed a range of distinct grammar types in order to capture different classes of linguistic capability. Though the essential aspects of human language still elude this approach, Chomsky's classification scheme was a boon for the study of computer languages and various types of computational device. What Young and I did was turn Chomsky's analogy between linguistic- and scientific-theory building inside out. We viewed the goal of a scientist as extracting from experimental data the linguistic structure of natural processes. This differs from *pattern recognition* in which data is compared against a pre-existing palette of patterns. Moreover, ours is not a qualitative approach but a quantitative one.

We developed a procedure—ϵ-machine reconstruction—to automate the discovery of grammatical rules hidden in experimental data. The rules were the "significant" patterns or regularities that govern the process which produced the data and that could be used to develop optimal predictions. The collection of the rules so discovered forms a "theory" of the process, in the sense that they model its mechanisms and allow us to make predictions about behavior that has yet to be observed.

In several ways, ϵ-machine reconstruction is analogous to a procedure, introduced by Norman Packard, Doyne Farmer, Rob Shaw, and myself, in 1980 for transforming experimental data into a geometric view of the "strange attractors" underlying deterministic chaos. In this light, the work with McNamara showed how this geometric approach could be extended to produce compact symbolic equations that governed the behavior on the attractors.

One fallout of ϵ-machine reconstruction was a much more refined notion—the *statistical complexity*—of information processing structures found in nature. Just as the excess entropy is complementary to the entropy rate, the statistical complexity as a measure of computation is complementary to the algorithmic notions of randomness introduced by Andrei Kolmogorov and Gregory Chaitin [2, 17]. Roughly speaking, the statistical complexity measures the amount of memory in a process; while Kolmogorov and Chaitin's *algorithmic entropy rate* measures how random a process is, when viewed as a computer. Thus, there can be a range of (structurally distinct) processes that each appears to be equally unpredictable, but that use different amounts of memory to produce that apparent randomness.

Young and I also introduced a useful graphical device—the complexity-entropy diagram—that reveals the range of information processing that natural systems can exhibit [9]. The complexity-entropy diagram is analogous to the thermodynamic *phase diagrams* introduced in the nineteenth century to map out the states of matter—solid, liquid, gas—at different conditions of temperature, pressure, and volume. It's different, though, in that it is based not on varying physical parameters, but on information processing coordinates: the rate at which information is produced (entropy rate) and how much memory is used to produce it (statistical complexity).

When we analyzed the boundaries between chaotic and predictable systems, we realized that the analogy with nineteenth-century thermodynamics was deeper than we had first thought. Just as water changes state in going from ice to liquid with increasing temperature, certain classes of information processing systems show phase transitions between order and chaos. The ordered regime is analogous to a crystalline solid; it literally corresponds to fixed crystalline patterns in time (periodic behaviors). The chaotic regime is analogous to a gas, in which the molecular motion is much more disordered. We demonstrated that at a order-chaos phase transition a new and qualitatively more powerful type of computation appears [10].

While different classes of natural process have their own computational-phase diagrams, our work suggested there are universal laws governing the interplay of the entropy rate and statistical complexity. It also indicated that there is organization at a higher level of understanding than the accounting of energy flows typically done in physics: the level of how natural systems store and process information and perform computations. Curiously, this view of the increase of complexity at the onset of chaos says, in a self-reflexive way, something more about the process of building scientific theories [5, 6].

4 THEORY

A key modeling dichotomy that runs throughout all of science is that between order and randomness. Imagine a scientist in the laboratory confronted after days of hard work with the results of a recent experiment—summarized prosaically as a simple numerical recording of instrument responses. The question arises, What fraction of the particular numerical value of each datum confirms or denies the hypothesis being tested and how much is essentially irrelevant information, merely "noise" or "error"?

This dichotomy is probably clearest within science, but it is not restricted to it, being a constant presence in the creation of artworks or in the engineering of artificial systems: What part of what we see or design is meaningful or functional? In many ways, this caricature of scientific investigation—"artificial science"?—gives a framework for understanding the necessary balance between order and randomness that appears whenever there is an "observer" trying to detect structure or pattern in its environment. The general puzzle of discovery then is: Which part of a measurement series does an observer ascribe to "randomness" and which part to "order" and "predictability?" Aren't we all in our daily activities to one extent or another "scientists" trying to ferret out the usable from the unusable information in our lives?

Given this basic dichotomy one can then ask: How does an observer actually make the distinction? The answer requires understanding how an observer models data—that is, the method by which elements in a representation, a "model," are justified in terms of given data.

A fundamental point is that *any* act of modeling makes a distinction between data that is accounted for—the ordered part—and data that is not described—the apparently random part. However, where to draw the line between theory and error is not so clear. The problem of building too complicated a model to fit all those things you want to explain is a familiar one in science. Jorge Luis Borges, the Argentine writer, illustrates the pitfall of "overfitting" in a faux critique of a nonexistent *Celestial Emporium of Benevolent Knowledge* [1], thusly:

> On those remote pages it is written that animals are divided into (a) those that belong to the Emperor, (b) embalmed ones, (c) those that are trained, (d) suckling pigs, (e) mermaids, (f) fabulous ones, (g) stray dogs, (h) those that are included in this classification, (i) those that tremble as if they were mad, (j) innumerable ones, (k) those drawn with a very fine camel's brush hair, (l) others, (m) those that have just broken a flower vase, (n) those that resemble flies from a distance.

As a general theory of "animal" the *Celestial Emporium* strikes us as being, at some points, too general and, at others, far too specialized, including too much "noise." Even without being a trained zoologist, one suspects that when presented with a candidate "animal" previously unknown to us, the *Celestial Emporium* may very well not help us in deciding whether or not it is an animal. As a scheme it does not generalize very well.

In principle, a balance between order and randomness can be reached and used to define a "best" model for a given data set. A balance can be found by minimizing the model's size while simultaneously minimizing the amount of apparent randomness or error. The first part is a version of Ockham's dictum [22]: causes should not be multiplied beyond necessity. The second part is a basic tenet of science: obtain the best prediction of nature. Neither component of this balance can be minimized alone, otherwise absurd "best" models would be selected. Minimizing the model size alone leads to huge error, since the smallest (null) model captures no regularities—all of the data appears to be noise; minimizing the error alone produces a huge model, which is simply the data itself and manifestly not a useful encapsulation of what happened in the laboratory. So both model size and the induced error must be minimized together in selecting a "best" model. Typically, the sum of the model size and the total error are minimized [24].

From the viewpoint of scientific methodology the key element missing in this view of modeling is how to measure structure or regularity. Just how structure is measured determines where the order-randomness border is set. This particular problem can be solved in principle: we take the size of the candidate model as the measure of structure. Then the size of the "best" model is a measure of the data's intrinsic structure. If we believe the data is a faithful representation of the raw behavior of the underlying process, this then translates into a measure of structure in the natural phenomenon originally studied.

After a little reflection one realizes, though, that this does not really solve the problem of quantifying structure. In fact, it simply elevates it to a higher level of abstraction. Measuring structure as the length of the description of the "best" model assumes one has chosen a language in which to describe models. The catch is that this representation choice builds in its own biases. In a given language some regularities can be compactly described, in other languages the same regularities can be quite baroquely expressed. For example, on the one hand, it is well known that, sentence for sentence, the German language expression of a thought is longer than the English equivalent. On the other, the sentiment captured in the single German word "freudenschade" has no equivalent in English and is translated to the longer phrase "happiness at other's distress". Change the language and the same regularities can require more or less description. And so, given that there is no prior God-given knowledge of the appropriate language for nature, a measure of structure in terms of the description length is, at root, arbitrary.

And so we are left with a deep puzzle, one that precedes measuring structure: How is structure discovered in the first place? If the scientist knows beforehand the appropriate representation for an experiment's possible behaviors, then the amount of that kind of structure can be extracted from the data as outlined above. In this case, the prior knowledge about the structure is verified by the data if a compact, predictive model results. But what if it is not verified? What if the hypothesized structure is simply not appropriate? Perhaps we've started out our data analysis with the wrong assumptions, the wrong representation. The "best" model could be huge or, worse, appear upon closer and closer analysis to diverge in size. The standard example of this is the Fourier—or frequency or sinusoidal or periodic—representation of the on-off "square wave". The Fourier representation describes the square wave as consisting of an infinite number of active frequencies; when, in fact, the square wave is described quite compactly (and exactly) as a "half on, half off" signal. The situation of an infinitely large model is clearly not tolerable. For one thing, it is impractical to manipulate. These situations indicate that the behavior is so new as to not fit (finitely) into current understanding. Then what do we do?

This is the problem of *innovation*. How can an observer ever break out of inadequate model classes and discover appropriate ones? How can incorrect assumptions be changed? How is anything new ever discovered, if it must always be expressed in the current language? If the problem of innovation can be solved, then, as all of the preceding development indicated, there is a framework which specifies how to be quantitative in detecting and measuring structure. One approach to this problem is hierarchical ϵ-machine reconstruction [5]. In this, one starts with the simplest assumptions about the world and then builds a succession of more sophisticated languages as the assumptions prove inadequate. ϵ-Machine reconstruction plays a central role in this because we use it to discover regularities, not in the raw data, but in a series of increasingly accurate models. Thus, we replace the data stream with a "model stream" and the regularities discovered form the basis of a new language that describes how less-accurate models are transformed into more-accurate ones.

5 CONCLUSION: THE MIDDLE GROUND

Copernicus said that the earth is not the center of the universe; Freud believed that our conscious self is the tip of an unknowable psychological iceberg. Gödel proved that there are limits to logical analysis; Turing, that answers can be beyond our reach; Poincaré that determinism leads to unpredictability; and Heisenberg that physical determinism fails on short temporal and small spatial scales.

The beautiful irony is that the result of each one of these concessions is an appreciation that the natural world is richer; that it is more structurally complex than we had previously thought. As individuals and as a culture we seem to be continually in a self-generated illusory state: saddled with implicit and naive assumptions about our ability to understand and control nature. These assumptions are only effective by dint of coincidence—in the sense that they are **not** nature, only feeble reflections of it. One might be tempted to view intellectual history as unkind, a continuing stripping away of these illusions. On retrospect, though, with each new fall, new knowledge and new understanding emerges.

Stepping back a bit, we now know that complexity arises in the middle ground, at the onset of chaos—the order-disorder border. Natural systems that evolve with and learn from interaction with their immediate environment exhibit both structural order and dynamical chaos. Order is the foundation of communication between elements at any level of organization, whether that refers to a population of neurons, bees, or humans. For an organism order is the distillation of regularities abstracted from observations. An organism's very form is a functional manifestation of its ancestor's evolutionary and its own developmental memory.

A completely ordered universe, however, would be dead. Chaos is necessary for life. Behavioral diversity, to take an example, is fundamental to an organism's survival. No organism can model the environment in its entirety. Approximation becomes essential to any system with finite resources. Chaos, as we now understand it, is the dynamical mechanism by which nature develops constrained and useful randomness. And from it follow diversity and the ability to anticipate the uncertain future.

There is a tendency, whose laws we dimly comprehend, for natural systems to balance order and chaos, to move to the interface between predictability and uncertainty. The result is increased complexity. This often appears as a change in a system's computational capability. The present state of evolutionary progress suggests that one need go even further and postulate a force that drives in time toward successively more sophisticated and qualitatively different computation. We can look back to times in which there were no systems that attempted to model themselves, as we do now. This is certainly one of the outstanding puzzles: How can lifeless and disorganized matter exhibit such a drive? And the question goes to the heart of many disciplines, ranging from philosophy and cognitive science to evolutionary and developmental biology and particle astrophysics. The dynamics of chaos, the appearance of pattern and organization, and the complexity quantified by computation will be inseparable components in its resolution.

Are these considerations too abstract to apply to contemporary social issues? I think not. At the very minimum, in a mathematical setting, understanding the interaction of order and chaos and the resulting complexity gives us a powerful set of metaphors for understanding more complicated (possibly complex) systems, such as human culture. In his *Process and Reality* [28], Whitehead saw a rather similar dynamic at work:

> The social history of mankind exhibits great organizations in their alternating functions of conditions for progress, and of contrivances for stunting humanity. The history of the Mediterranean lands, and of western Europe, is the history of the blessing and the curse of political organizations, of religious organizations, of schemes of thought, of social agencies for large purposes. The moment of dominance, prayed for, worked for, sacrificed for, by generations of the noblest spirits, marks the turning point where the blessing passes into the curse. Some new principle of refreshment is required. The art of progress is to preserve order amid change, and to preserve change amid order. Life refuses to be embalmed alive. The more prolonged the halt in some unrelieved system of order, the greater the crash of the dead society.

Can we as individuals come to appreciate the dynamic balance of order and chaos? Will our societies self-organize into a dynamic that moves beyond the least common denominator results characteristic of human groupings, toward an organization that is appreciative of diversity, understands the role of regularity, and that is truly and constructively complex? Economies, the scientific community, international relations, and other societal groupings are extremely large, complicated systems. Nonetheless, in the more limited and abstract realm of mathematics and physics we are beginning to see some glimmers of order amid the chaos, to appreciate the constructive role of randomness, and to understand the dynamic interplay of order and chaos. What lies between order and chaos? The answer now seems remarkably simple: Human innovation. The novelist and lepidopterist Vladimir Nabokov appreciated more deeply, than many, the origins of creativity in this middle, human ground [20]:

> There is, it would seem, in the dimensional scale of the world a kind of delicate meeting place between imagination and knowledge, a point, arrived at by diminishing large things and enlarging small ones, that is intrinsically artistic.

REFERENCES

[1] Borges, J. L. *Other Inquisitions 1937–1952.* New York: Simon and Schuster, 1964.

[2] Chaitin, G. "On the Length of Programs for Computing Finite Binary Sequences." *J. ACM* **13** (1966): 145.

[3] Chaitin, G. "Randomness in Arithmetic." *Sci. Am.* **July** (1988): 80.

[4] Chomsky, N. "Three Models for the Description of Language." *IRE Trans. Info. Th.* **2** (1956): 113.

[5] Crutchfield, J. P. "The Calculi of Emergence: Computation, Dynamics, and Induction." *Physica D* **75** (1994): 11–54.

[6] Crutchfield, J. P. "Is Anything Ever New? Considering emergence." In *Complexity: Metaphors, Models, and Reality*, edited by G. Cowan, D. Pines, and D. Melzner, 479–497. Santa Fe Institute Studies in the Sciences of Complexity, Proc. Vol. XIX. Reading, MA: Addison-Wesley, 1994.

[7] Crutchfield, J. P., and B. S. McNamara. "Equations of Motion from a Data Series." *Complex Systems* **1** (1987): 417–452.

[8] Crutchfield, J. P., and N. H. Packard. "Symbolic Dynamics of Noisy Chaos." *Physica* **7D** (1983): 201–223.

[9] Crutchfield, J. P., and K. Young. "Inferring Statistical Complexity." *Phys. Rev. Let.* **63** (1989): 105–108.

[10] Crutchfield, J. P., and K. Young. "Computation at the Onset of Chaos." In *Entropy, Complexity, and the Physics of Information*, edited by W. Zurek, 223–269. Santa Fe Institute Studies in the Sciences of Complexity, Proc. Vol. VIII. Reading, MA: Addison-Wesley, 1990.

[11] Derrida, J. *Of Grammatology.* Baltimore: Johns Hopkins University Press, 1976.

[12] Freud, S. *Civilization and Its Discontents.* New York: W. W. Norton, 1961.

[13] Freud, S. "Determinism – Chance – And Superstitious Beliefs" in *The Basic Writings of Sigmund Freud.* New York: Modern library, 1995.

[14] Gopnik, A., A. N. Meltzoff, and P. K. Kuhl. *The Scientist in the Crib: Minds, Brains, and How Children Learn.* New York: William Morrow and Company, 1999.

[15] Heisenberg, W. *The Physical Principles of the Quantum Theory.* Chicago: The University of Chicago Press, 1930.

[16] Kolmogorov, A. N. "A New Metric Invariant of Transient Dynamical Systems and Automorphisms in Lebesgue Spaces." *Dokl. Akad. Nauk. SSSR* **119** (1958): 861.

[17] Kolmogorov, A. N. "Three Approaches to the Concept of the Amount of Information." *Prob. Info. Trans.* **1** (1965): 1.

[18] Levi-Strauss, C. *Triste Tropiques.* New York: Atheneum, 1973.

[19] Lorenz, E. N. "Deterministic Nonperiodic Flow." *J. Atmos. Sci.* **20** (1963): 130.

[20] Nabokov, V. V. *Speak Memory: An Autobiography Revisited.* New York: Everymans Library, 1999.

[21] Nagel, E., and J. R. Newman. *Gödel's Proof.* New York: New York University Press, 1968.

[22] Ockham, William of. *Philosophical Writings: A Selection, Translated, with an Introduction, by Philotheus Boehner, O.F.M., Late Professor of Philosophy, The Franciscan Institute.* Indianapolis: Bobbs-Merrill, 1964.

[23] Poincaré, H. *Les Methodes Nouvelles de la Mecanique Celeste.* Paris: Gauthier-Villars, 1892.

[24] Rissanen, J. *Stochastic Complexity in Statistical Inquiry.* Singapore: World Scientific, 1989.

[25] Shannon, C. E., and W. Weaver. *The Mathematical Theory of Communication.* Champaign-Urbana, IL: University of Illinois Press, 1962.

[26] Turing, A. M. "On Computable Numbers, with an Application to the Entsheidungs Problem." *Proc. Lond. Math. Soc. Ser. 2* **42** (1936): 230.

[27] van der Pol, B., and J. van der Mark. "Frequency Demultiplication." *Nature* **120** (1927): 363.

[28] Whitehead, A. N. *Process and Reality.* New York: The Free Press, 1978.

[29] Zurek, W., ed. *Entropy, Complexity, and the Physics of Information.* Santa Fe Institute Studies in the Sciences of Complexity, Proc. Vol. VIII. Reading, MA: Addison-Wesley, 1990.

Art and Complexity
J. Casti and A. Karlqvist (editors)

Regularities and Randomness: Evolving Schemata in Science and the Arts

Murray Gell-Mann

It is a pleasure to be back in Abisko and to participate in this discussion of simplicity and complexity in science and in the arts. At the Santa Fe Institute, which I helped to found and where I now work, we devote ourselves to studying, from many different points of view, the transdisciplinary subject that includes the meanings of simplicity and complexity, the ways in which complexity arises from fundamental simplicity, and the behavior of complex adaptive systems, along with the features that distinguish them from non-adaptive systems.

My name for that subject is *plectics*, derived from the Greek word *plektós* for "twisted" or "braided," cognate with *–plexus* in Latin *complexus*, originally "braided together," from which the English word complexity is derived. The word *plektós* is also related, more distantly, to *plex* in Latin *simplex*, originally "once folded," which gave rise to the English word *simplicity*. The name *plectics* thus reflects the fact that we are dealing with both simplicity and complexity.

I believe my task today is to throw some light on plectics and to indicate briefly how it illuminates parallels between certain processes associated with the arts and certain phenomena studied in the sciences or characteristic of the scientific enterprise itself. We can begin with questions such as these:

What do we usually mean by complexity?

What is chaos?

What is a complex adaptive system?

Why is there a tendency for more and more complex entities to appear as time goes on?

It would take a number of quantities, differently defined, to cover all our intuitive notions of the meaning of complexity and of its opposite, simplicity. Also, each quantity would be somewhat context dependent. In other words, complexity, however defined, is not entirely an intrinsic property of the entity described; it also depends to some extent on who or what is doing the describing. (There is one exception to this context dependence, encountered, for example in the mathematical theory of computational complexity, when one considers a sequence of similar systems of larger and larger size and looks only at their behavior as the size approaches infinity.)

Let us start with a rather naïvely defined quantity, which I call "crude complexity"—the length of the shortest message describing the entity. First of all, we would have to exclude pointing at the entity or calling it by a special name; something that is obviously very complex could be given a short nickname like Sam or Judy, but giving it that name would not make it simple. Next, we must understand that crude complexity will depend on the level of detail at which the entity is being described, what we call in physics the coarse graining. Also, the language employed will affect the minimum length of the description. That minimum length will depend, too, on the knowledge and understanding of the world that is assumed: the description of a rhinoceros can be abbreviated if it is already known what a mammal is. (Imagine how long it would take to explain to a recently contacted Amazonian Indian what is meant by a tax-managed mutual fund.)

Having listed these various kinds of context dependence, we can concentrate on the main feature of crude complexity, that it refers to length of the shortest message. In my book, *The Quark and the Jaguar*, I tell the story of the elementary school teacher who assigned to her class a three hundred word essay, to be written over the weekend, on any topic. One pupil did what I used to do as a child—he spent the weekend poking around outdoors and then scribbled something hastily on Monday morning. Here is what he wrote:

> "Yesterday the neighbors had a fire in their kitchen and I leaned out of the window and yelled 'Fire! Fire! Fire! Fire!...'" If he had not had to comply with the three hundred word requirement, he could have written instead "...I leaned out of the window and yelled 'Fire!' 282 times." It is this notion of *compression* that is crucial.

Now in place of crude complexity we can consider a more technically defined quantity, algorithmic information content. An entity is described at a given level of detail, in a given language, assuming a given knowledge and understanding of the world, and the description is reduced by coding in some standard manner to a string of bits (zeroes and ones). We then consider all programs that will cause a standard universal computer to print out that string of bits and then stop computing. The length of the shortest such program is called the algorithmic information content (AIC). This is a well-known quantity introduced over thirty years ago by the famous Russian mathematician Kolmogorov and by two Americans, Gregory Chaitin and Ray Solomonoff, all working independently. We see, by the way, that it involves some additional context dependence through the choice of the coding procedure and of the universal computer. Because of the context dependence, AIC is most useful for comparison between two strings at least one of which has a large value of it.

A string consisting of the first two million bits of the number pi has a low AIC because it is highly compressible: the shortest program just has to give a prescription for calculating pi and ask that the string be cut off after two million entries. But many long strings of bits are incompressible. For those strings, the shortest program is one that lists the whole string and tells the machine to print it out and then halt. Thus, for a given length of string, an incompressible one has the largest possible AIC. Such a string is called a "random" one, and accordingly the quantity AIC is sometimes called algorithmic randomness.

We can now see why the AIC of the entity being described does not correspond very well to what we usually mean by its complexity. Compare a play by Shakespeare with the typical product, of equal length, of the proverbial ape at the typewriter, who types every letter with equal probability. The AIC, or algorithmic randomness, of the latter is overwhelmingly likely to be much greater than that of the former. But it is absurd to say that the ape has produced something more complex than the work of Shakespeare. Randomness is not what we mean by complexity.

Instead, let us define what I call *effective complexity*, the AIC of the *regularities* of an entity, as opposed to its incidental features. A random (incompressible) bit string has no regularities (except its length) and thus very little effective complexity. Likewise something extremely regular, such as a bit string consisting entirely of ones, will also have little effective complexity, because its regularities can be described very briefly. To achieve high effective complexity, an entity must have intermediate AIC and obey a set of rules requiring a long description. But that is just what we mean when we say that the grammar of a certain language is complex, or that a certain conglomerate corporation is a complex organization, or that the plot of a novel is very complex—we mean that the description of the regularities takes a long time. The same is true of the U.S. tax code, or of Japanese culture.

The famous computer scientist, psychologist, and economist Herbert Simon used to call attention to the path of an ant, which has a high AIC and appears complex at first sight. But when we realize that the ant is following a rather simple program, into which are fed the incidental features of the landscape and the pheromone trails laid down by the other ants for the transport of food, we understand that the path is fundamentally not very complex. Herb says, "I got a lot of mileage out of that ant." And now it is helping me to illustrate the difference between crude and effective complexity.

When we discuss the shortest program that will produce the regularities of an entity, it is not only the length of that program (the AIC of the regularities) that matters. Also important is the length of time (or the number of steps) that it takes the standard computer, starting with a short program, to print out a description of the regularities and then stop computing. That length of time is called the logical depth of the regularities. Sometimes we fail to recognize that an entity has regularities that are governed by a short program, so that the effective complexity looks high when it is really low. When that happens, we may possibly be dealing with a high value of logical depth, so that the short program requires a very long calculation to yield the regularities. This situation is often encountered in theoretical science. For instance the energy levels of atomic nuclei look complicated at first, as if their regularities would require a very long description. But we now believe they are governed by a very simple and elegant theory. However, the calculations necessary to go from that theory to the prediction of the energy levels are too lengthy for existing computers to carry out! Assuming the theory is right, the energy levels have very little effective complexity, but lots of logical depth.

There can be no finite procedure that is guaranteed to find all the regularities of an entity. We may ask, then, what kinds of things engage in identifying sets of regularities. The answer is: complex adaptive systems, including all living organisms on Earth.

A complex adaptive system receives a stream of data about itself and its surroundings. In that stream, it identifies particular regularities and compresses them into a concise "schema," one of many possible ones related by mutation or substitution. In the presence of further data from the stream, the schema can supply descriptions of certain aspects of the real world, predictions of events that are to happen in the real world, and prescriptions for behavior of the complex adaptive system in the real world. In all these cases, there are real world consequences: the descriptions can turn out to be more accurate or less accurate, the predictions can turn out to be more reliable or less reliable, and the prescriptions for behavior can turn out to lead to favorable or unfavorable outcomes. All these consequences then feed back to exert "selection pressures" on the competition among various schemata, so that there is a strong tendency for more successful schemata to survive and for less successful ones to disappear or at least to be demoted in some sense.

Take the human scientific enterprise as an example. The schemata are theories. A theory in science compresses into a brief law (say a set of equations) the regularities in a vast, even indefinitely large body of data. Maxwell's equations, for instance, yield the electric and magnetic fields in any region of the universe if the special circumstances there—electric charges and currents and boundary conditions—are specified. (We see how the schema plus additional information from the data stream leads to a description or prediction.)

In biological evolution, the schemata are genotypes. The genotype, together with all the additional information supplied by incidents in the process of development—in the case of higher animals, from the sperm and egg to the adult organism—determines the character, the "phenotype," of the individual adult. Survival to adulthood of that individual, sexual selection, and success or failure in producing surviving progeny all exert selection pressures on the competition of genotypes, since they affect the transmission to future generations of genotypes resembling that of the individual in question.

In the case of societal evolution, the schemata consist of laws, customs, myths, traditions, and so forth. The pieces of such a schema are often called "memes," a term introduced by Richard Dawkins by analogy with genes in the case of biological evolution.

For a business firm, strategies and practices form the schemata. In the presence of day-to-day events, a schema affects the success of the firm, as measured by return to the stockholders in the form of dividends and share prices. The results feed back to affect whether the schema is retained or a different one substituted (often under a new CEO).

A complex adaptive system (CAS) may be an integral part of another CAS, or it may be a loose aggregation of complex adaptive systems, forming a composite CAS. Thus a CAS has a tendency to give rise to others.

My colleague John Holland uses a different terminology. What I call a CAS is something like what he calls an adaptive agent. He reserves the term "complex adaptive system" for a composite CAS, consisting of agents that are adapting to one another, such as organisms in an ecological system or investors in a market. What I call a schema he calls an internal model. We are both illustrating the famous principle that a scientist would rather use someone else's toothbrush than another scientist's nomenclature.

On Earth, all complex adaptive systems seem to have some connection with life. To begin with, there was the set of prebiotic chemical reactions that gave rise to the earliest life. Then the process of biological evolution, as we have indicated, is an example of a CAS. Likewise each living organism is a CAS. In a mammal, such as a human being, the immune system is a complex adaptive system too. Its operation is something like that of biological evolution, but on a much faster time scale. (If it took hundreds of thousands of years for us to develop antibodies to invading microbes, we would be in serious trouble.) The process of learning and thinking in a human individual is also a complex adaptive system. In fact, the term "schema" is taken from psychology, where it refers to a pattern used by the mind to grasp an aspect of reality. As we have seen, aggregations of human beings can also be complex adaptive systems: societies, business firms, the scientific enterprise, and so forth.

Nowadays, we have computer-based complex adaptive systems, such as "neural nets" and "genetic algorithms." While they may sometimes involve new, dedicated hardware, they are usually implemented on conventional hardware with special software. Their only direct connection with life is that they were developed by human beings. Once they are put into operation, they can, for example, invent new strategies for winning at games, strategies that no human being has discovered.

Science fiction writers and others may speculate that in the distant future a new kind of complex adaptive system might be created, a truly composite human being, by wiring together the brains of a number of people. They would communicate not through language, which (according to an aphorism attributed to Voltaire) is used by men to conceal their thoughts, but rather through sharing all their mental processes. My friend Shirley Hufstedler says she would not recommend this procedure to couples about to be married.

The behavior of a complex adaptive system, with its variable schemata undergoing evolution through selection pressures from the real world, may be contrasted with "simple" or "direct" adaptation, which does not involve a variable schema, but utilizes instead a fixed pattern of response to external changes. A good example of direct adaptation is the operation of a thermostat, which simply turns on the heat when the temperature falls below a fixed value and turns it off when the temperature rises above the same value.

In the study of a human organization, such as a tribal society or a business firm, one may encounter at least three different levels of adaptation, on three different time scales.

1. On a short time scale, we may see a prevailing schema prescribing that the organization react to particular external changes in specified ways; as long as that schema is fixed, we are dealing with direct adaptation.
2. On a longer time scale, the real world consequences of a prevailing schema (in the presence of events that take place) exert selection pressures on the competition of schemata and may result in the replacement of one schema by another.
3. On a still longer time scale, we may witness the disappearance of some organizations and the survival of others, in a Darwinian process. The evolution of schemata was inadequate in the former cases, but adequate in the latter cases, to cope with the changes in circumstances.

It is worth making the elementary point about the existence of these levels of adaptation because they are often confused with one another. As an example of the three levels, we might imagine a prehistoric society in the U.S. Southwest that had the custom (1) of moving to higher elevations in times of unusual heat and drought. In the event of failure of this pattern, the society might try alternative schemata (2) such as planting different crops or constructing an irrigation system using water from far away. In the event of failure of all the schemata that are tried, the society may disappear (3), say with some members dying and the rest dispersed among other societies that survive. We see that in many cases failure to cope can be viewed in terms of the evolutionary process not being able to keep pace with change.

Individual human beings in a large organization or society must be treated by the historical sciences as playing a dual role. To some extent they can be regarded statistically, as units in a system. But in many cases a particular person must be treated as an individual, with a personal influence on history. Those historians who tolerate discussion of contingent history (meaning counterfactual histories in addition to the history we experience) have long argued about the extent to which broad historical forces eventually "heal" many of the changes caused by the acts of individuals.

A history of the U.S. Constitutional Convention of 1787 may make much of the conflicting interests of small states and large states, slave states and free states, debtors and creditors, agricultural and urban populations, and so forth. But the compromises invented by particular individuals and the role that such individuals played in the eventual ratification of the Constitution would also be stressed. The outcome could have been very different if certain particular people had died in an epidemic just before the Convention, even though the big issues would have been the same.

How do we think about alternative histories? Is the notion of alternative histories a fundamental concept?

The fundamental laws of nature are:

1. the dynamical law of the elementary particles—the building blocks of all matter—along with their interactions,
2. the initial condition of the universe near the beginning of its expansion some thirteen billion years ago.

Theoretical physicists seem to be approaching a real understanding of the first of these laws, as well as gaining some inklings about the second one. It looks as if both may be rather simple and knowable, but even if we learn what they are, that would not permit us, even in principle, to calculate the history of the universe. The reason is that fundamental theory is probabilistic in character (contrary to what one might have thought a century ago). The theory, even if perfectly known, predicts not one history of the universe but probabilities for a huge array of alternative histories, which we may conceive as forming a branching tree, with probabilities at all the branchings. In a short story by the great Argentine writer Jorge Luis Borges, a character creates a model of these branching histories in the form of a garden of forking paths.

The particular history we experience is co-determined, then, by the fundamental laws and by an inconceivably long sequence of chance events, each of which could turn out in various ways. This fundamental indeterminacy is exacerbated for any observer—or set of observers, such as the human race—by ignorance of the outcomes of most of the chance events that have already occurred, since only a very limited set of observations is available. Any observer sees only an extremely coarse-grained history.

The phenomenon of *chaos* in certain nonlinear systems is a very sensitive dependence of the outcome of a process on tiny details of what happened earlier. When chaos is present, it still further amplifies the indeterminacy we have been discussing.

A few years ago, at the wonderful science museum in Barcelona, I saw an exhibit that beautifully illustrated chaos. A nonlinear version of a pendulum was set up so that the visitor could hold the bob and start it out in a chosen position and with a chosen velocity. One could then watch the subsequent motion, which was also recorded with a pen on a sheet of paper. The visitor was then invited to seize the bob again and try to imitate exactly the previous initial position and velocity. No matter how carefully that was done, the subsequent motion was quite different from what it was the first time. Comparing the records on paper confirmed the difference in a striking way.

I asked the museum director what the two men were doing who were standing in a corner watching us. He replied, "Oh, those are two Dutchmen waiting to take away the chaos." Apparently, the exhibit was about to be dismantled and taken to Amsterdam. But I have wondered ever since whether the services of those two Dutchmen would not be in great demand across the globe, by organizations that wanted their chaos taken away.

Once we view alternative histories as forming a branching tree, with the history we experience co-determined by the fundamental laws and a huge number of accidents, we can ponder the accidents that gave rise to the people assembled in this room. A fluctuation many billions of years ago produced our galaxy, and it was followed by the accidents that contributed to the formation of the solar system, including the planet Earth. Then there were the accidents that led to the appearance of the first life on this planet, and the very many additional accidents that, along with natural selection, have shaped the course of biological evolution, including the characteristics of our own subspecies, which we call, somewhat optimistically, Homo sapiens sapiens. Finally we may consider the accidents of genetics and sexual selection that helped to produce the genotypes of all the individuals here, and the accidents in the womb, in childhood, and since that have helped to make us what we are today.

Now most accidents in the history of the universe don't make much difference to the coarse-grained histories with which we are concerned. If two oxygen molecules in the atmosphere collide and then go off in one pair of directions or another, it usually makes little difference. But the fluctuation that produced our galaxy, while it too may have been insignificant on a cosmic scale, was of enormous importance to anything *in* our galaxy. Some of us call such a chance event a "frozen accident." Once it has occurred, it can be responsible for a good deal of regularity.

I like to quote an example from human history. When Arthur, the elder brother of King Henry VIII of England, died—no doubt of some quantum fluctuation—early in the sixteenth century, Henry replaced Arthur as heir to the throne and as the husband of Catherine of Aragón. That accident influenced the way the Church of England separated from the Roman Catholic Church (although the separation itself might have occurred anyway) and changed the history of the English and then the British monarchy, all the way down to the days of Charles and Diana.

It is the frozen accidents, along with the fundamental laws, that give rise to regularities and thus to effective complexity. Since the fundamental laws are believed to be simple, it is mainly the frozen accidents that are responsible for effective complexity. We can relate that fact to the tendency for more and more complex entities to appear as time goes on.

Of course there is no rule that everything must increase in complexity. Any individual entity may increase or decrease in effective complexity or stay the same. When an organism dies or a civilization dies out, it suffers a dramatic decrease in complexity. But still the envelope of effective complexity keeps getting pushed out, as more and more complex things arise.

The reason is that as time goes on frozen accidents can keep accumulating, and so more and more effective complexity is possible, provided the rate of accumulation of the consequences of frozen accidents outstrips the rate at which such consequences die out. This phenomenon occurs even for non-adaptive evolution, as in galaxies, stars, planets, rocks, and so forth. It is, of course, well known in biological evolution, where in some cases higher effective complexity probably confers an advantage. And we see all around us the appearance of more and more complex regulations, instruments, computer software packages, and so forth, even though many things become simplified.

The tendency of more and more complex forms to appear in no way contradicts the famous second law of thermodynamics, which states that for a closed (isolated) system, the average disorder ("entropy") keeps increasing. There is nothing in the second law to prevent local order from increasing, through various mechanisms of self-organization, at the expense of greater disorder elsewhere. (One simple and widespread mechanism of self-organization on a cosmic scale is provided by gravitation, which has caused material to condense into the familiar structures with which astronomy is concerned, including our own planet.)

Here on Earth, once it was formed, systems of increasing effective complexity have arisen as a consequence of the physical evolution of the planet over some four and half billion years, as well as biological evolution over four billion years or so. On a much shorter time scale, human cultural evolution has also given rise to things of greater and greater complexity. In all these cases, it is the accumulation of the results of frozen accidents that has allowed the proliferation of regularities to take place.

Now every complex adaptive system picks out certain regularities and compresses their description into schemata. But how does a complex adaptive system identify regularities?

First of all, we have to recognize that a CAS can make mistakes in identifying regularities. We human beings make such errors all the time. In our search for comforting order in a universe that depends a great deal on chance, we often discover regularities that aren't there. That is a good way to describe superstitions. We also frequently engage in denial, ignoring regularities that are staring us in the face.

When a CAS discovers genuine regularities, it typically uses mutual information as a diagnostic. We can define mutual information for two or more bit strings as the information that the strings contain in common. Now suppose the information available to a CAS about an entity is processed in some way into a bit string and that string is somehow divided into parts. Then a large amount of mutual information among those parts is diagnostic of regularities, provided the string is not redundant and the process of creating it from the available data is such that it does not introduce false regularities. (In more technical language, we can say, concerning the information available about the entity, that a high degree of internal mutual AIC is diagnostic of regularities.) Thus the CAS discovers regularities of an entity by looking for internal similarities in the information about it. That is certainly what a person does in looking at a work of art or listening to a piece of music or reading a work of literature.

When a CAS finds in an entity regularities "of interest" to it and formulates a schema describing those regularities, the length of the schema will in general be quite different from the minimum program length defining the effective complexity of the entity. Assume the regularities are genuine. They will still usually be incomplete, making for a shorter schema, except that much underlying simplicity may be missed, making for a longer schema. In addition, the schema will typically be somewhat redundant, for convenience and also to facilitate error correction, and that will make it longer than it has to be.

Once again, it is important to emphasize that the available information about the entity depends very much on the level of detail at which it is described and also on the knowledge and understanding of the world that is assumed. If a person looks at a painting from a very great distance, the resulting very coarse graining means that only gross features of the painting will be used in finding regularities in it and thus judging its complexity. Also, a considerable number of devices used by the painter will be wasted on a viewer who does not have the requisite experience. To take a trivial example, recall the famous single line tracing the ears and back and tail of a cat. That very simple abstraction immediately suggests a cat, with many appropriate connotations, but only to people who are already familiar with the animal in question. A work of minimalist art may contain very few bits, especially if seen from afar, but if it affects the viewer and draws upon his or her store of experience, it can convey far more information.

In comparing the arts and the sciences, perhaps the most striking differences lie in the nature of the regularities with which the two enterprises are concerned. The poet, for example, may find and emphasize regularities in the world that have little to do with scientific theory, but depend on associations of ideas and images deep in the human mind. Again, some degree of community of experience between the writer and the reader or listener may be necessary in order for those associations to be appreciated.

One point of resemblance between the sciences and the arts lies in the importance of symmetry, and especially broken symmetry, to the schemata in both areas. Symmetry represents, of course, one of the most striking forms of regularity.

But nearly everything we experience exhibits a mixture of regularity and randomness, or regularity combined with features that are treated as incidental. Take, for example, a male bird of a given species singing a territorial song in the nesting season. Listening to that bird, or to many birds of the same species, one can spot regularities in the sample of song and also variations that seem to be random or incidental. The listener may be a scientist, finding regularities in the hope of figuring out what characteristics of the song are really necessary in order to frighten off other males or attract a suitable female and thus define a territorial song of this species.

A powerful method of describing identified regularities consists of embedding the observed sample in a very large set of otherwise imaginary samples, with a probability attached each member of the set. A set equipped with probabilities is called an *ensemble.* The ensemble in question reflects the regularities. The observed sample should be a typical member of that ensemble, in the sense that its probability should not be unusually low.

It is worth remarking that the description of the ensemble can be much shorter than a description of every one of the individual members along with its probability. For example, we can speak of the set of all human beings alive today without specifying all their names and addresses.

To summarize: whether or not the entity being observed by the CAS is itself a set of comparable things, the genuine regularities identified by the CAS in that entity can be described by embedding it in an imaginary ensemble. The entity is then treated *conceptually* as a typical member of an ensemble of entities that share the identified regularities, while the other members of the ensemble illustrate the possible individual variations compatible with those regularities. If the description of that set is short, then the CAS is attributing a low effective complexity to the entity. If the minimal description appears to be very long, then the CAS is attributing a high effective complexity to it. But, of course, to the extent that the CAS may have overlooked a simpler description of the regularities, that high value of the complexity may be wrong.

The notion of placing the imagined alongside the real in order to get insight into the nature of the real is critical both in the sciences and in the arts. In classical statistical mechanics, we may be dealing with a sample of gas with a trillion trillion molecules, each with an initial position and velocity. That amount of information about the sample is impossible to gather or to store or to utilize. Instead, we embed that sample conceptually in an ensemble of samples, all the rest imagined. That ensemble may be characterized by just a few parameters, such as temperature, that help us to understand the regularities of the original gas sample.

In a somewhat analogous way, the fiction writer or the dramatist supplies us with imaginary situations and characters to place alongside the real ones we encounter in life. When the work is well done, that juxtaposition allows us to appreciate better the nature of real people and their relationships. In poetry and the visual arts, we can see much the same phenomenon in a more subtle form. As our experience is enlarged by exposure to the arts, we see the world around us in new ways.

Now in discussing the arts, we may adopt either the point of view of the artist or else that of a viewer, listener, or reader (say, "viewer" for short). In the former case, we are concerned with the evolution of artistic schemata as prescriptions for creation, whether in the generation of a single work, in the development of an artist's oeuvre, or in the history of a school or movement. In the latter case, we are dealing with the schema used by an external viewer to describe the work. In each situation, as we have emphasized, the information about the work depends on the coarse graining and on the context, and in each case a division is made between the regular and the random.

In the case of the individual work, the regularities can be described by embedding it in a conceptual ensemble. For the oeuvre or the work of the school, we may describe the regularities by embedding the whole series of pieces in a conceptual ensemble of series. Throughout, we are considering the artist as a CAS, the school or movement as a loose aggregation of complex adaptive systems functioning more or less as a CAS, and the viewer as a CAS learning about the art in question.

The selection pressures on the artist or the school include internal conceptions of what the art should be like; external pressures from critics, the market, and viewers in general; social and political pressures from the community at large; and the usual pressures on someone learning by making mistakes and then correcting them on the basis of further experience. Some of the selection pressures on the viewer have similar origins. To a greater or a lesser extent, depending on the situation, there is co-evolution between the artist or the school on the one hand and the buyers, funders, critics, and ordinary viewers on the other. They educate one another.

As selection pressures feed back on the competition among the relevant schemata, evolution takes place and we can picture the evolution of a schema in terms of an ensemble, a conceptual cloud or swarm moving in an abstract space, typically with a higher density toward the middle and much lower densities near the edges. As in the earlier discussion, a description of the cloud is a description of the perceived regularities, while the spread of the cloud represents the scope of individual variation allowed by those regularities. The information content in the description of the cloud is the effective complexity attributed by the schema, while the average additional information content in the examples, once the description of the cloud is given, is the random information attributed by the schema.

It is worth remarking that in contemporary semantics the meaning of a word or expression is no longer treated so much as a problem of delineating sharply between what is meant and what is not, as in some dictionary definitions. Instead, it is recognized that there is likely to be a central meaning (or perhaps more than one) around which a cloud of secondary meanings stretches in various directions in "meaning space," with decreasing weight as the distance from the central meaning increases. As the usage of the word evolves through time, the cloud moves and the weights shift.

It is fascinating to trace the evolution of art styles conceived in this way, especially when we are dealing with a succession of real sets of examples. Coinage and pottery are two areas that lend themselves particularly well to this kind of study. In the days of hammered coinage, before the advent of the milling machine, there was a great deal of individual variation within a given issue of coins. After a wave of barbarian conquests, the scope of such variation would increase, and the types would generally become cruder as well. After the incursions that finished off the Roman Empire in the West, coin types started to move away from the imperial models. Not only did the imperial portrait and other symbols evolve, but also the legends were altered, sometimes to the point of becoming meaningless.

We can think of the regularities in the coins issued by a given set of related mints around a given time in terms of an ensemble, a cloud in an abstract coin space. After many decades of evolution into more barbaric forms, one can see in some cases a condensation of the cloud around a new model. A fantastic animal may appear, for example, in place of an old portrait, or a new legend may arise referring to a barbarian king instead of the Byzantine emperor. Either the evolution of the regularities as seen by an outside observer or the evolution of the corresponding schemata in the various mints can be represented in terms of the motion of a cloud.

In the production of an individual work of art, the amount of evolution that takes place in the artist's schema during the creation of a work seems to vary a good deal from artist to artist. One artist may claim that he or she prepares a plan and then follows it, with only small changes in the course of the creation. Another may describe a tortuous process of trial and error, involving lots of twists and turns, with selection pressures coming from artistic judgment, the possibilities of the medium, and perhaps comments from other people.

One interesting feature of the regularities in the whole oeuvre of an artist comes up in the process of distinguishing fakes from the real thing, especially in the visual arts. A highly skilled faker may be able to copy the most conspicuous features of the artist's style and even deceive many experts, but a minor characteristic that is nevertheless a true regularity may escape the faker's attention and give him away if an expert becomes aware of it. Carlo Ginzburg, in a recent essay on the subject, cites the authors who first pointed this out and quotes some of their examples, such as the shape of a fingernail in the work of a particular painter. In studying the arts, it often pays to track down true regularities, even if at first their significance is not obvious. The same holds true for the sciences. Nature exhibits regularities, and, as Isaac Newton emphasized, "It is the business of natural philosophy to find them out."

Art and Complexity
J. Casti and A. Karlqvist (editors)

Drawing, Knowledge, and Intuitive Thinking: Drawing as a Way to Understand and Solve Complex Problems

Bobo Hjort

A question of interest today, exemplified by the theme of the seminars is how scientists should approach art. I will here suggest that scientists should start drawing, and what I thus want to emphasize is that a scientist has little benefit of meeting with artists compared to what he can reach by learning and practicing *the method* of artists'.

As an architect my profession exists somewhere between science and art, which can mean that architects are neither scientists nor artists, or that they are both. Although most architects are not specialists in either of these two fields, they are possibly what the Norwegian philosopher Arne Kvaløi has named "supermateurs," indicating that they belong to both fields. They are at least not exaggeratedly respectful of either of them, which I believe is important.

Tor Nørretranders has talked about the importance of seeing the complexity of everyday life. I will propose that we look at art, in all its complexity, as an every day phenomenon, since by taking a servile or detached attitude to art it is impossible to learn from it. By describing how an architect works, as a link between the two fields, I will try to show how scientists can learn from, and make better use of the artist's way of thinking.

When I receive a commission I start immediately to produce drawings, projections, and pictures. People often consider drawings as illustrations, the *result* of a mental effort, but what I want to discuss is drawing as a process, including all the drawings of which more than 95% are for my eyes only. I don't start by analyzing the situation, I am not even sure that I read the program thoroughly. I just start to draw, and soon after I have received the commission I have a first sketch on the drawing table. It is not only a part of the building or a detail. It is a solution to the total problem, or an idea about a solution. It does not at all solve the demands of the program, but the aim is to create a representation of the whole.

From the moment I have this drawing I can start the next phase, which is to examine it. I look at it, view it, but not very carefully. I don't examine how it solves the program; I consider it as a picture. I am not interested in finding faults, only in how to go forward. I think I look for something that has to be changed or developed. Exactly how this works I do not know, but suddenly I am working with the next sketch. When I then observe that one, I get new impulses to change and start new sketches and in this way I go on, until I am satisfied. That does not mean that my job is finished, just that I now have a grasp on it, and do not feel the need to continue any longer.

I have probably established a form of problem hierarchy, and I can go on with the demands of the program and elaborate on details, parts, and new points of view. This is because I have caught the totality in a form of an idea or a structure to which I can relate them. Of course, the details influence the whole, and I have to move between different levels, but all the time in the same manner: Draw, observe, draw, observe, which means much trial and silent, aesthetic evaluation.

I have considered the possibility of creating buildings in a more intellectual or calculating way, by analyzing the problem, dividing it into smaller problems, solving each at a time and then to try to coordinate them. I have doubts about this method however, since its difficulties increase in proportion to the degree of complexity, and to create a building is, in my view, a very complex task. The architect's method has obvious advantages, so I myself continue to draw and have now started to explore the possibility that it might be useful to other professional problem solvers.

The work of an architect includes of course a lot of ordinary analytic thinking, but I will leave it out in this text since I want to focus attention exclusively on the non-analytic or intuitive activity, and specifically the following components.

1. You have to start. This is a naïve comment, but important because it says that you have to be brave, and that you must not listen to the voice that tells you to wait for a better idea.
2. The process will not follow a pattern of hierarchy or casual connection. I myself don't know when I start what will happen next. It is first when I *look* at my sketches I *feel* what to do.
3. The fundamental mental work is to draw and to evaluate *without words.* It is not a logic verbal analysis but an intuitive immediate judgement of problems and possibilities.
4. My solution is not an answer to the given program or question. The program changes and deepens during the process, parallel to the development of the solution. I probably start with an answer, and then try to adjust that answer and the question to each other.
5. I do not stop because I have found *the* solution; I stop because I lose interest or the urge to continue. My interpretation of this phenomenon is that I have unconsciously gained insight to a problem, and formulated it as a solution. The drawing is formulated knowledge.

The main words in this process are *drawing* and *intuition.*

Drawings are indispensable because they more easily, and clearer than words describe an imagined building, but they have other interesting characteristics.

1. One that is often mentioned is that a picture constitutes a whole. This means that you can quite easily detect different mistakes like undesired consequences or impossible connections.
2. Another characteristic is that a picture can never be the result of a calculation, or be the correct answer to a given question. What it can be is a commentary on a problem, a *spontaneous and subjective commentary.*
3. A picture can not be the final solution to a problem but a possible solution.
4. Its aim is not *the* truth, but it can be true.
5. A drawing can not be judged or criticized only logically. The judgement passed must be subjective, which means that it has to be based on values existing on a level where words do not reach. I believe that pictures and drawings appeal to deep human knowledge and experiences.

Drawing can be a way to get in contact with silent or wordless knowledge connected to experience and a profound value system, and thus start a mental process capable of handling complex problems without a definite solution. In other words, drawing uses another kind of thinking than the logical/verbal. It uses *visual/intuitive* thinking.

Intuition, in my words, is a mental system that tells us how to act. It is older than analytic and causal thinking, and I believe that early man was, as animals are, directed by intuition. I know that some people are suspicious of intuition, but I claim that we make many decisions daily with its help. We dress, buy appropriate food and choose the bicycle instead of the car on a sunny day. It would be a waste of resources to make more use of our intellectual capacity in these situations.

Even very important decisions are based on intuition, and that to a higher degree than most people want to believe. A managing director must trust his intuition when he makes important decisions. If he constantly needs more facts he is of no use. When we choose a partner for the rest of our lives we very seldom try to analyze our situation, and draw conclusions, we prefer to believe in our intuition. Intuition is not a whimsical impulse. It is a message from our deepest well of knowledge.

I argue that we have two modes of thinking; one logical/verbal, which I dare call the scientist's way, and one intuitive/wordless, which I consequently call the artist's way. I consider that we all use both, but that we do not pay the artist's way enough respect, even though we are all too respectful to art.

So in my opinion we do not exploit the capacity of the intuitive way of thinking, and consequently mismanage our own resources. I do not say that we should *always* rely on intuition, but that intuitive thinking can help us to understand and solve complex problems. Design theory tells us that it is not possible to solve a design task with only intellectual thinking, and I believe that many problems can be understood as design tasks.

There are or course more methods to reach, develop, and exploit one's intuitive thinking than by making drawings, and many people have probably discovered their own, but still I will point out some advantages of the one I practice.

1. To create a mental vision or picture is a common way for most people to understand words. Especially abstract notions.
2. All of us have practiced drawing for many years when we were children. Drawing was once a natural way for us to examine, understand, and describe.
3. We can all use it as soon as we stop setting up rules and principles for what the pictures *should* look like.
4. Drawing is a way to see and understand because you have to observe carefully and from different positions. Drawing forces you to observe.
5. Drawing is the only method to develop intuitive thinking, that I know, which is traditionally taught at a university level. In my department, as in all architectural departments we teach the sketch method.

The sketch method is continuously trained in the projects, but the basic skills are taught in drawing-, painting-, and sculputure-courses. The aims of these courses are that the student shall

1. learn not to be ashamed of their drawings. This is very important and can take a long time.
2. discover that their pictures show something that they did not know they had seen.
3. discover that their drawings contain knowledge that they did not already know.
4. discover the links between their inner images and their pictures.
5. discover the links between prejudgments and creative thinking.

What I have tried to describe and discuss is one method to develop and use one's intuitive thinking, and my question is if this method can be useful to other professions than architects. In order test this, our department with Ylva Dahlman as teacher in charge, has been offering courses in drawing to students in other disciplines; veterinarians, agronomists, engineers. We very clearly declare that our aim is not to make artists of them, and not to teach them techniques, but to help them to understand their own subjects in a richer way. After four years the courses seem, according to the evaluation, to be successful for many students, and I therefore believe that they could be useful even for scientists who want to develop their visual, intuitive thinking. My hope is that they, as the students, shall discover that drawing is a good way to comprehend and solve complex problems.

Art and Complexity
J. Casti and A. Karlqvist (editors)

Nothing is Hidden

Anders Karlqvist

1 KNOWLEDGE AND COMPLEXITY

A typical feature of our "information society" is that knowledge is encoded in a way that allow storing, manipulating, and transferring by the modern machine—the computer. The efficiency of computerization hinges on the seemingly obvious fact that the computer makes it possible to mange complexity. Technological progress goes hand in hand with more complex applications and more sophisticated expressions of human creativity as demonstrated in the success story of science.

Computerization is only the latest phase in the history of knowledge management and computer technology only one spectacular tool in a series of inventions, where the most revolutionary steps probably were taken in the dawn of human history with the development of speech and later with the invention of written language, the latter a most important landmark in knowledge and hence complexity management.

It is possible to discern these revolutionary steps as particular instances of a general trend in the history of information technologies and society's management of knowledge. This trend can be characterized as a process of *externalization* in the following sense: The internal and implicit structure of a certain phenomena is formalized and made explicit. Hence it is given an form which can be communicated and manipulated. Thereby it provides a model which can be extended and applied to new situations and generate new patterns.

Language is a basic tool for expressing human experiences. It is based on a set of rules which especially in the written language becomes quite well defined. We can now express and keep record of vast amount of information. A notational system with letters or symbols provides the tool kit which is applied in a standardized way.

This means typically a shift of complexity. When we go from an oral tradition to written language complex sounds are replaced by simple symbols like letters which can be combined in new ways to produce words and sentences of great complexity. Codification in written form also provides new possibilities to store and transmit knowledge. Structure becomes more important and more attention has to be given to grammar and syntax. In this sense computer language represents an extreme case of elementary simplicity (with the symbols 0 and 1) and with almost unlimited possibilities for combinatorial complexity.

What is gained and what is lost in such a codification process? It is a nontrivial question and this aspect becomes particularly relevant in artistic work. Often art tries to capture "the residual" of what cannot be represented in a formalized scientific mode. The problem of representing complex phenomena with simple standardized building blocks and explicit sets of rules is then highlighted. The field of music provides some illustrative examples how these difficulties arise.

2 COMPLEXITY IN MUSIC

Historically, music has its root in oral practice and traditions have developed based on direct communication between musicians. Little attention has given to the role of the composer as distinguished from the performer. The written notation for music emerged in Western society probably in ancient Greece but did not develop into an effective system for polyphonic music until the eleventh century. It took another couple of hundred years before also rhythm and time values became properly formalized and the modern musical score could develop in the form which is well known today. With this notational system which is most typically used in classical and contemporary western music it has become possible to preserve and transfer the knowledge about music to other places and times and liberate the musical product from the producer, the composer. The note system is simple, clear and precise. This precision has had an overwhelming influence on a flourishing musical culture in Western society.

In the early days, say before the eighteenth century and still today in certain traditional music and non-western music the tonal scale, the pitch and the tuning of instrument may vary in many ways. When the notational system was introduced and providing effective means of exploring the full range of tonalities it became urgent to standardize the tuning of instruments. This was especially important for instruments where the intervals between notes were given and could not be manipulated by the player (like a piano as opposed to e.g., a violin).

The remarkable set of Preludes and Fugues, das Wohltemperierte Klavier, by J. S. Bach (the first 24 were finished in 1722) marks this transition to a well tempered tuning with equal intervals between all notes in the scale. It is a compromise making all keys sounding equally "false." Some of the special sounds in music which is based on other tonalities and which would be preserved by oral tradition is lost through the notation but on the other hand it opens up for an effective use of polyphonic music and complex works such as nineteenth century symphonies for big orchestras.

It has been argued that the efficieny of this notational system is not an all together positive contribution to the art of music. The musical expression becomes standardized in a way that obscure certain qualities, qualities which are part of the tacit knowledge of musical performance. This tacit knowledge is essentially of two kinds: Knowledge which is taken for granted in a specific culture of music, such as note inégale in baroque music or certain rythmic conventions in jazz and on the other hand knowledge which cannot be described or articulated in the notational systems, such as slight differences in pitch or intonation.

One might argue that it would be possible to refine the notational system to compensate for such deficiencies, i.e., to increase the precision making the system more elaborate, thereby embracing more of the musical complexity. However this give rise to other objections such as how to account for improvisation and flexibility. Music is more than imitation. It is an act of creation and in many sorts of music, not only in jazz, improvisation plays an extremely important role. It is an integral part of musical knowledge. It is human-dependent tacit knowledge which cannot be completely formalized. It is not difficult to distinguish between music generated by a machine (such as the melodies programmed into our mobile telephones) and music performed by a live musician.

So how can we make the notational system better? Or should we refrain from making it better? In an empirical study of Norwegian folk music the musicologist Henrik Sinding-Larsen discovered that one considered good replicas of a tune (e.g., by music scores or records) as more threatening than the bad ones. The living tradition of folk music would not benefit from increased perfection in this sense. A good description generally means a context-free description which means the folk musicians as a social group (the context) lose control over their music tradition. A lesson to be learnt from this example is that notation which is more precise, more standardized and a more efficient tool to provide an external representation of (in this case musical) knowledge, does not automatically eliminate the task of interpreting and expressing complexity. It rather shifts the focus. Music remains a balancing act between externalized and tacit knowledge.

3 FROM DESCRIPTION TO PRESCRIPTION

Another outcome of such an externalization process is a possible shift from description to prescription. What was originally intended as a method to describe music with symbols becomes a method of deriving musical expressions within the given frame set by the notational system. Music becomes more or less "defined" as what can be expressed within this system and its limits set by the notes and the score. The descriptive tools become objects of description themselves and hence part of an co-evolution of means and ends.

Externalization introduces the possibility of a meta-level. The tools become objects. The language serves as a tool for studying the language itself. This kind of self-referencing plays an important role in the science of complexity and has profound philosophical as well as scientific implications. When John von Neumann was thinking about how to construct a self-reproducing automata he made the crucial observation that information is treated in two different ways, as raw data and as instructions. In computer parlance we can say that information can be on one hand interpreted as a program and on the other treated as input data (for a program to act on). This turns out to be exactly the mechanisms which rule the reproduction of living beings where the DNAs (unknown to von Neumann) have the role of carrying the information. If we see reproduction as a kind of externalization (making a copy of oneself), complexity through self-reference is the crucial step needed and a necessary ingredient for life.

Self-reference is also the source of logical paradoxes. The problem with statements such as "this statement is false" was recognized already in ancient Greece. A similar statement was used by Kurt Gödel in his famous proof that a mathematical system powerful enough to include arithmetic is not complete. In other words there are statements in mathematics which are true but cannot be proven. The basic trick is to construct a sentence in mathematical language which says: "this statement cannot be proven." What Gödel managed to do was to derive a mathematical formula which was at the same time a statement *in* mathematics and *about* mathematics.

Gödel's theorem and other related results have been used as an argument and inspired claims that human knowledge cannot be formalized. The human mind is not computable, not even in principle. This is not the place to elaborate on those ideas. There exists a vast literature for and against the computer metaphor in cognitive science. Let me just refer to a couple of arguments put forward by John Searle, which I think are relevant also in the broader context of complexity and formalization of knowledge. Why does the machine fail to pass the Turing test? In spite of all efforts to program a machine to respond intelligently to human questions its limitation as an expert system remains obvious. Searle gives the following example:

A man goes into a restaurant. Did he order a hamburger? The computer will answer yes or no (depending on other details of the story) and hence it seems to respond intelligently. But if you ask: Did the man eat his hamburger through his ear or mouth? Or, is the hamburger more than 30,000 years old? The computer would most likely answer: I don't know. The problem, as Searle describes it, is the difficulty of programming the background or the context or the common sense, which is an integral part of human understanding. This background does not consist of representations but rather in a set of abilities, stances, non-representational attitudes and general know-how. This background forms the precondition of intentionality but it is not itself intentional.

4 TO FOLLOW RULES

Another argument to see the difference between computers and humans deals with the idea of what it means to follow a rule. In the sense in which human beings follow rules, computers do not follow rules at all. They act only in accord with certain formal procedures. Searle points out that computers act only *as if* they are following rules. We use rule-following only in a metaphorical sense. This metaphor tends to be confused with the literal meaning of rule-following which is applicable to human behavior.

What do we mean with understanding a rule? For Ludvig Wittgenstein it meant to master the corresponding practices. It is through acting that words get their meaning. In his own words: "A rule in itself does not give any fundamental insight. What matters does not do so in virtue of some intrinsic and self-evident quality. It is rather held fast by what lies around it." This reminds of Searle's comment of the necessity to see the context or background. I deliberately use the word "see." We are taught how to *know* what to *see*. Wittgenstein insists that everything about human behavior is open to observe. Nothing is hidden. But to those who do not understand specific practices it will remain unintelligible.

Wittgenstein gave a series of lectures on the foundation of mathematics in Cambridge 1939. Among his small group of students was Alan Turing who became famous for breaking the German enigma code (in addition to his formulation of the Turing machine and work on gödelian results). He was also one of the pioneers in the emerging field of computer science. Turing's position was very much along the lines of what later was to be the AI-movement and later in life (he died in 1954) he wrote an article called: Can a machine think?

It is evident from the records of the lectures that Wittgenstein and Turing held opposite views regarding the nature of mathematics. For Wittgenstein mathematics was not a matter of discovery but of inventing and using concepts according to certain practices, practices established through an agreed usage. To say that 3 times 7 is 22 is to not follow the rules. Nothing more or less. Logical paradoxes which Russell and other philosophers pondered over were in Wittgenstein's view dead ends and waste of time.

To actually do mathematics or to understand mathematics has much less to do with explicit rule-following than one would be inclined to believe. Mathematics provides no doubt a very precise and explicit language for making logical deductions and deciding what is true and what is not. However, truth itself is not a very good guide. There are many true statements in mathematics which are totally irrelevant. Relevance is a product of the background and the cultural context. The famous case of the Indian genius Ramanujan is a striking example. He had a brilliant mathematical mind and established great many theorems about numbers but he had almost no contact with the mathematical community and a broader mathematical context. His contributions, although extremely intriguing, remain largely a curiosity.

5 TRUTHS AND LIES

Francis Bacon noted that truth emerges more readily from error than from confusion. In the proof of Fermat's last theorem a crucial step relies on a deep result (about elliptical curves and modular forms) by Shimura and Taniyama. Shimura speaks in admiration about his colleague Taniyama: "He was gifted with the special capability of making many mistakes mostly in the right direction."

So let me suggest that "True or false" is a blunt criteria for scientific understanding. We may note that very few of all possible mathematical propositions and true statements mean anything and add to real understanding. We need to summon all our mental powers, both intuitive and aesthetic. Actually false statements can sometimes be more interesting and revealing than truth, an observation which is well recognized in the field of the arts. Pablo Picasso asserted that Art is a lie which makes us see the truth, at least such truths that we able to grasp.

In art as well as in science it is worth recognizing the importance of vagueness and ambiguity as a source of creativity. A mathematical calculation obeys strictly logical principles and this is exactly what is needed to tell a machine in a computer program. As we all know such descriptions are tedious and almost incomprehensible to follow for a human mind. We want to know not what needs to be done according to logic but why. We need to relate to a wider context to understand and appreciate. We use pictures and metaphors. As Banach told his fellow mathematician Stan Ulam: "Good mathematicians see analogies between theorems. The very best ones see analogies between analogies."

6 TACIT KNOWLEDGE

"Nothing is hidden. The difficulty is to see what lies openly before our eyes." As a student in mathematics I was confronted with a book in topology by Nicolas Bourbaki. There was a statement in the preface which I think Wittgenstein himself would have liked. It went something like this: "This book does not require any prior knowledge of mathematics only a certain measure of mathematical maturity." It is not a very encouraging declaration to guide a young student in his studies. It gives, however, an interesting message about the essence of mathematical knowledge. You are supposed to master the tacit knowledge and the idea of mathematical thinking. The text will only provide you with a rigorous description of all facts.

By the way, there is an irony related to this recommendation. N. Bourbaki does not exist as a physical person. It was a pseudonym for a group of French mathematicians, which was kept more or less anonymous. So here we have a virtual scientist commenting on tacit knowledge!.

Concepts can function as powerful cognitive tools with the background information which they entail. Sometimes concepts can be extended beyond their original meaning and in ways not intended. DNA-molecules are interpreted as letters and words, reactions in particle physics as arrows in Feynman diagrams and so on. By building associations to well known phenomena which are easy to grasp, symbols and signs not only become an economic method of storing and manipulating knowledge but also serve as a source of inspiration to think in new creative directions. The elegance of e.g., Einstein's field equations in general relativity is partly illusive since the symbols and the formulas contain a tremendous amount of physical and mathematical knowledge which can be unfolded, understood and used only by he who masters this background. This is a common situation in science as well as in the arts. The more effective we become in expressing knowledge in notational systems of various kind and hence in storing and transmitting knowledge effortlessly, the greater is the risk for a widening gap between accessible knowledge and the ability to use it. This gap needs to be explored by scientists as well as by artists

Art and Complexity
J. Casti and A. Karlqvist (editors)

Science and Art in Collaboration—The Mindship Method

Tor Nørretranders

Science has a lot to learn from art. And vice versa. An experience in collaboration, and the principles behind, are presented. Art will in the future become a Gold mine for industry, like science has been in the past.

1 THE COMPLEXITY OF EVERYDAY LIFE

In 1990, I noticed an odd detail in a painting. At the time I was a guest lecturer teaching the modern scientific world view at the Schools of Visual Arts of the Royal Danish Academy of Fine Arts in Copenhagen.

The odd detail was a tree-like structure appearing in paintings done by a student there, Malene Bach. The trees looked a little fractal to me, but then again not at all.

At the time, pattern formation was a major issue in science, mostly thanks to the advent of chaos theory and nonlinear mathematics as well as the revival of the geometrical approach to nature brought about by Benoit Mandelbrot and his campaign for the fractal geometry of nature. So I was eager to know which of the different new approaches to the physics of patterns that would explain this. I discussed it with a physicist friend, Søren Brunak, and later Jakob Bohr. Both of them were astonished by the phenomenon. We decided to study it.

"Why do you want to do that?," said Malene Bach, the art student producing the tree-like structures in her paintings. "It's trivial. Every artist knows this. It's just a well-known trick," she said. "Max Ernst used it, many people use it: You take two canvasses, sandwich some paint in between them, draw them apart and the paint will produce a double tree-like structure, one on each canvas."

So this was trivial to artists. But not to scientists. We looked in the literature and could only locate one reference to the phenomenon, giving a wrong explanation for it.

We tried then in a hobby-like fashion to reproduce the phenomenon in different fluids. And it turns out to be extremely robust: Sandwich in between two plates a blot of paint, mayonnaise, toothpaste, hair grease, butter—just any viscous fluid. When the plates are separated: Whoops, there you have two trees, each on a plate, mirror-images of each other. The specific form is never repeated, not two trees are identical, every time the shape is different. We called them *viscous trees.*

The phenomenon is very easily reproduced: My favorite liquid is toothpaste sandwiched between two flat surfaces found in a kitchen (plates, pot lids, etc.).

What is this, what is the mechanism behind, we asked ourselves. It quickly turned out that it was a very special case of a well-studied phenomenon, much in focus in those early days of experiments into chaos theory. The phenomenon is called Saffmann-Taylor fingering and describes how an interface between air and fluid will behave when forces act upon it. In a way, viscous tress comprised the simplest and most robust example of the experimental production of a phenomenon from the world of chaos in the kitchen—using only ingredients present in any household.

In further studies, we were able to simulate the growth of the trees. In experimental studies, we could show that the trees resulted from a novel mechanism of pattern formation, competition for space between fingers of air invading the footprint of a liquid on a plate [2].

All this was a lot of fun and even a little worthwhile scientific work. But the artists didn't care. The phenomenon was trivial to them. Later, I have noticed the tree structure in many other painters' work. So it is trivial, yes, but it was not understood in terms of science.

Which brings me to my first point: The world of art harbors an enormous tacit knowledge of how the world functions, lots of insight into very specific phenomena in e.g., the behavior of fluids, unstudied by science.

This is the case for many professions: The study of Saffman-Taylor fingering started in the 1950'ies when scientists were alerted by problems in the printing industry. In many walks of life, people are familiar with phenomena of the real world that scientists do not know of, or choose to ignore because they are too complicated.

The computer and the science of fractals/chaos have provided scientists with a little more self-confidence in addressing the issues raised by non-scientists: Why are clouds of the shape they are, how come trees look like they do, why are mountains peaks upon peaks?

Science has a long tradition of ignoring the real world. And in particular the questions that children ask about the real world. And for good reasons. Science has had absolutely nothing intelligent to say about these questions that children ask. Perhaps old heroes like Leonardo tried to study trees or eddies, but since then, these problems were simply to complicated to be dealt with through the help of Newtonian mathematics and physics. So everyday reality was left to itself, while scientists constructed their science and their technological devices.

But thanks to the kids, we are constantly reminded that we cannot explain why the immediate world has taken the form it has. The arts help too. And the computer and the nonlinear revolution has helped scientists acknowledge that here was indeed something one could study in a meaningful way. So my first point is this: Science needs art. Scientists should study all the tacit knowledge of the material world that is present in the arts. There's a bag of tricks there, known to any arts student, but begging scientific explanation.

This general point that science needs art and that scientists should study the folklore of the arts world, is not, of course, limited to the science of materials, such as the intricacies of air-fluid interfaces.

Art is a way of knowing the world and getting to know the world. Artists have an enormous bank of knowledge about materials, light, optics, human perception, visual illusions, the interpretation of space, interaction between objects and attention—not to speak of all the psychological knowledge present in the world of fiction.

In general, everyday life is full of phenomena that are too complicated for scientists, yet essential to everyone else. Not only viscous trees, but the interplay between human beings, the weather, the tricks of cooking, the infinite universe of female beauty. All these are important realities, many of them left untouched by science. Because they are too complicated.

Chaos theory taught us something very essential: Even if we know the laws of nature, like Newton's laws of motion, we still do not know what nature will look like. Knowing the laws does not allow us to predict how they will in fact turn out to produce phenomena. It takes more information to describe natural phenomena than is present in the laws of nature. They are very abstract and cannot tell us how the concrete world will look.

That doesn't mean that real world in not governed by the laws of nature. It means that just because you can dissect the world into causes and effects, and just because you can understand the laws connecting cause and effect, it doesn't mean than you can reconstruct this world.

This point was brought out many years ago by the solid state physicist P. W. Anderson in a visionary lecture from 1967 [1].[1] But now the world of chaos has made it clear to everyone. Chaos theory showed us that even fully determined systems, totally governed by rigid laws, could be impossible to predict, even in principle.

Its' like saying: Just because the world is made up of atoms, and any material object can be dissolved into individual atoms (for instance by heating it into a gas), it doesn't mean that starting with a few zillion atoms and a textbook of atomic physics, you can deduce how an elephant sucks up water.

So if we want to understand the world, it is not enough to study the laws of it, like the scientists prefer to. We also have to study the world itself, like for instance they way artists do.

(An analogous example is this: Just because almost all of world literature can be said to describe the problems involved in pairing human beings of opposite sex so that they can have fun and reproduce—and this is very true of most fiction, in

[1]The article is based on a lecture from 1967 and further discussed in Tor Nørretranders [4]

fact—it doesn't mean that you can reconstruct the latest novel or movie based on this insight.)

So scientists need art and artists. Perhaps they can even bring something worthwhile with them into the world of art: A systematic language, a lot of theory, a lot of human courage and curiosity, a lot of technology and know-how and a lot of the poetry of science.

Science and art need each other.

2 BRINGING PEOPLE TOGETHER

In 1996, when Copenhagen was the Cultural Capital of Europe, we created *ad hoc* the Mindship to host a series of international seminars between artists and scientists.[2]

The starting point was the need for collaboration between the two worlds and the strongly held view that enough seminars had been dealing with art-meets-science at some general level of abstraction.

The object was not to unite or relate the two worlds in general (for one reason, if not many other, because nobody can really explain what science or art are in general, but all practitioners know very well what a good work of science or art is like, once they see it).

The aim was to select a number of very specific, concrete and everyday-like phenomena that was being explored simultaneously in the world of art and the world of science, but without the two necessarily being in contact. We chose issues like case studies in the reproduction of mental states (psychology-medicine/theatre-music) or cases of perception of complexity and beauty (mathematics-biology/music-visual arts).

The seminars were organized such that 15 scientists met 15 artists over a three week period. During each of the 15 working days of the three weeks, a scientist and a artist were given a chance to present their work. In the morning, a scientist, in the afternoon, an artist. Hence, participants were at the same time students and teachers: For instance teaching their art to the scientists and learning science from the same people. Thereby, each group acted as a proxy for the general public. When artists explain the traditions of improvised music to scientists they have to assume the same level of ignorance in their audience, as when they approach the general public. They cannot use musical notation etc.

Therefore, the two groups of professionals had to communicate in terms of very down-to-earth concrete language, not some technical and abstracts models. To further ensure that the two groups were not lost in abstractions leading nowhere, we asked them to go on stage twice a week, presenting their work to an audience of lay copenhageners who had bought a ticket for the evening.

Thus, we insisted that the language of the seminars should be everyday language. Not jargon. This meant that the scientists had to explain to themselves what their scientific theories meant, just like the artists had to explain themselves, freed of the traditions of the art world. For both groups this created some very strong emotional reactions, both negative and positive.

[2]Some results are stored at ⟨http://www.mindship.org⟩.

The interaction with the public served as an attractor, keeping the meeting between the professions safely anchored in the complexity and wilderness of real life, rather than the abstracts laws of nature and the tacit nodding of know-it-all art critics.

Apart from this pragmatic motivation, our point was to emphasize that any concrete situation in life can be labeled and described in many more ways than the people involved will ever do. If you are in a situation, you can never ever describe that situation in full detail, unless you are willing to let the situation slip away. (You are standing there with the girl you want, it's a beautiful sunset at the beach, and you start explaining the details of the weather, of the refraction of the light rays from the Sun in the atmosphere, the number of grains of sand visible on the beach, the locomotion of crabs, the hum of insects, her hair, the wind, your low back pain, problems with your wet feet—for heaven's sake: kiss her, right now, sweetheart.)

There are always room for many, many different levels of description in almost any situation. It could be a novel, a treatise in atmospheric physics, a study of body language, whatever. The world is rich. There is always room for both science and art in the full description of just about any situation in this universe.

As physicist Freeman Dyson so poetically expressed it (with a little help from an old German friend of his): The universe is infinite in all directions [3].

Pick any direction and it's fruitful. Yet many academics and art snobs tend to mistake their little universe of abstraction and tacit connotations for the real world in all its chaotic richness.

We need to learn to co-exist with other professions than our own, dealing with this richness. There is plenty of room for all.

And yes, the Mindship seminars turned out to be extremely fruitful. We provided our guests with plenty of time, space, and opportunity to cooperate. We gave them very concrete ideas of issues on which they could cooperate, conjecturing how they could learn from each other. These conjectures, of course, turned out to be wrong, but in the process of shooting them down, collaboration had started without anyone noticing it.

Sadly, the 1996 tradition couldn't continue, but there was no doubt that it was a worthwhile thing to do.

3 THE GOLD MINE AHEAD

Artists have an enormous experience in dealing with the world. They often deal with issues that are ignored by the world of science. They are open to the fact that the world is complex and that human beings often perceive it in odd ways. One can argue (with Marshall McLuhan) that artists are the real experts in perception.

In the coming decades, the economies of the Western world will increasingly deal with communication, experience, entertainment, messages, visualizations, etc. etc. The real experts in all this are the artists. They know how human beings see the world and imagine stuff. They therefore have much to offer the world of software, games, human-machine interface, brain science, fluid dynamics, etc. etc.

In the twentieth century science successfully provided industry and state with the means to grow and win power. From the chemical industry to the nuclear programs of the Cold War, science was the basis of the execution of economic and military power in the world. For good and bad, science was the basis of civilization.

This was a great opportunity for science to grow and learn new stuff. It was also a temptation for it, to get all too well acquainted with the seductive forces of power and money. Science lost much of its virginity in the process, but then the budgets were growing all the time, so that during most of the Cold War science could exercise autonomy.

Only after the end of the Cold War and after the cuts in science budgets of the past ten years, do we now see the full extent of the erosion of science into commercial and institutional services.

The role of science has changed. Not necessarily to the better.

Similarly, in this twenty-first century, art will become the basis of the economy. No longer will it be engineering or behavior control science that will be of indispensable importance to industry and state. It will be the knowledge and folklore of film editors, musicians, painters, dancers, and illusionists that form the basis for the development of the entertainment industry, telecommunication, nanotechnology, and even biotechnology.

Who are we, where do we come from and how will we be entertained—those are the questions that will provide the basis for the future of economy.

Scientists may become envious of the wealth and power of artists to come, but they already know the prize: Loss of reality contact, loss of professional sovereignty, loss of fun and joy.

Making art and science work together is a worthwhile endeavor, if for no other reason because artists now need to learn some first-aid help from the scientists.

REFERENCES

[1] Anderson, P. W. "More is Different." *Science* **177** (1972): 393–396.

[2] Bohr, Jakob, Søren Brunak, and Tor Nørretranders. "Viscous Trees and Voronoi Structure Formation in Expanding Systems." *Europhys. Lett.* **25** (1994): 246–251.

[3] Dyson, Freeman. *Infinite in All Directions.* New York: Harper & Row, 1988.

[4] Nørretranders, Tor. *The User Illusion.* New York: Penguin, 1999.

Art and Complexity
J. Casti and A. Karlqvist (editors)

Complexity and Emergence in the American Experimental Music Tradition

Tim Perkis

I'm going to talk today a little bit about my own work, but primarily about a particular tradition, the American experimental music tradition which my generation of composers has inherited. This tradition has a strange and unique character, I think, which gives it a special relevance for our topic here: in short, the music coming out of this tradition is explicitly concerned with the perception and appreciation of complex dynamical systems. I do electronic music. In my case, this generally means that I set up a system of interacting components of some kind. Sometimes the piece consists of a computer program, sometimes it's a set of analog electronic equipment, sometimes it consists of systems that involve people, and instructions to people. It could be all three, or it could be a network of computers. Generally I design some process of interaction, and allow it to behave. This behavior is what makes the music—in fact this behavior *is* the music.

Of course I didn't invent the whole idea of working this way—there's a clear chain of development throughout this century that led to this practice. When I started doing this kind of work in the late 1970s, there was a very active scene of people working this way in the San Francisco Bay Area, many around Mills College in Oakland, which has long been a major center for new music in the United States. There was the very exciting feeling in the air that this was the way music was going to be made now, that we were on the threshold of a new way of thinking about things that was going to change culture in a major way. Unfortunately, I've found that while we were right about this to some degree—certainly using computers to make music, for example, has become something that nearly everyone accepts—this particular way of working has other aspects that are often misunderstood. And so now I feel that when I talk about my music I need to also talk about the historical context in which this way of working arose. It's actually an interesting story. I believe the American experimental music tradition is unique, arising from a strange confluence of things that has led to a way of thinking about music, and working in music that is unlike what goes on in other traditions. It has elements of the influence of science and of technology, of the visual arts, of Asian philosophy, of European music—especially French music, I think—and it also has a distinctive American kind of rebelliousness about it.

The salient feature of this tradition is its unique and characteristic way of thinking about the activity of making music. The music is seen not primarily as implementing a vision of the composer, or the will of the composer—something the composer hears in his head. Rather it's about setting up situations that allow the appearance of sonic entities that are more like natural phenomena than traditional music. The practitioners of this type of music build machines, or things akin to machines or simulations, things that have a behavior of some kind that is unanticipated by the composer.

The modernist American composer Morton Feldman once, when asked whether he hears the music in his head before he writes it, said, more or less "people who hear music in their head are nuts—they should be locked up. That's not what a composer does." In his view, what the composer does is set up a situation, set it in motion, and observe, listen. In essence, once that happens, the composer's position is not that different from the audience. He or she is capable of being as surprised as anyone by what actually happens in the music.

This is an unusual way to think about music, and there is a still very healthy, living alternate way to think about music, the more traditional view of what the composer does, of somehow pulling the music out of his head. Of course Beethoven provides the preeminent archetype of this conception: the lone genius (deaf, yet!) whose mind is full of completely realized symphonies, and who struggles to write fast enough to capture them. I suppose there are people like that—I don't know anyone like that—and in fact, that's not a way of thinking about music that particularly interests me, and has not been what this American tradition I'm discussing is about.

This heroic, romantic view of the composer, and of music as some sort of ectoplasmic excretion of a mind or soul, is actually not all that old, really arising in the 18th century and gaining pre-eminence in the 19th. There are pre-romantic antecedents of the American experimental perspective, in which music is seen as somehow more external: whether in the medieval conception of music as a divine visitation, or the late renaissance aesthetic of music as providing an image or representation of real physical phenomena. The role of the composer in the experimental view is in a sense more passive than that of a romantic composer: once set in motion, the music has its own life and the composer is a listener like any other. Calling this music experimental is quite precisely correct: like a scientist setting up an experiment, the experimental composer sets up the conditions that define the piece, and is interested in hearing what actually happens with it.

I'm now going to discuss some examples of different music I think belongs in this tradition. The first example is from someone who is the grandfather of the American experimental tradition, one who has had a great influence on many American composers: Charles Ives. There are a rich variety of innovations in Ives' music, but the selection I draw your attention to, from his *Symphony No. 4* (1916), conveys something of his experimental attitude. He described how as a child he went to a Fourth of July parade—a parade with more than one band in it—and as the bands went by, there would be moments when two bands could be heard at once. They were playing completely separate pieces, completely uncoordinated rhythmically, harmonically, in any way. In one movement of the *Symphony No. 4* he recreates that phenomenon with an orchestra.

This music is an attempt to represent something he heard, a depiction of an acoustic effect. The simultaneous sound of two bands is of course a man-made phenomenon, but Ives' interest in it is in its aspect as an uncontrolled, naturally occuring event. Representing this event in his symphony is not primarily an act of self-expression: it's really about listening, and exploring the world on its own terms.

Once one has decided that musical work in some sense involves studying the behavior of entities beyond oneself, than a next logical step is to actually construct situations that exhibit musical behavior. Rather than composing music, the composer designs an algorithm, a virtual machine, which he uses to generate a score for players. More recently, composers working this way will build literal machines, electronic circuits, or software machines that generate music. These machines are not really machines for a specific, well understood purpose like a car, or a watch: they really have more in common with a mathematical simulation. As with a simulation, what the composer is really interested in is eliciting some unknown behavior. One designs a machine, an algorithm, which is perhaps predictable, but the new relationships that arise in the musical product of this mechanism are unknown, in fact unknowable in advance. A pioneer in this approach to music in the early twentieeth century is Henry Cowell, a Californian. In 1930 he built a rhythm machine, with a big mechanically driven wheel, which held pegs set to play different rhythms, and one could study different rhythmic patterns this way.

John Cage, who is central to this entire tradition, was another musical bricoleur. Cage used to quote Arnold Schonberg's comment about him, that he was "not a composer, but an inventor—of genius." And this is quite accurate—he was an inventor of musical machinery of different types, machinery that has musical behavior, providing us with new and undreamed of musical sensations. Cage's 1947 work *Sonatas and Interludes* illustrates the machine-like aspect of his work well. This work is a series of short movements for prepared piano, a piano with various items stuck in the strings to make it a miniature percussion orchestra. Simple and strict geometric and arithmetic procedures are followed in the composition of the piece. While the conceptual framework is meticulously planned, the actual musical affect is not pre-conceived. Cage makes it quite explicit that the underlying intent of his work is a spiritual one, the idea of opening our ears to hear things we haven't heard before. He talks about escaping our own tastes, and escaping our cultural prejudices. Central to the American take on modernism in music is the notion that, like science, it's about the discovery and perception of alien, unknown phenomena.

I keep calling this an American tradition, and it is important to look at what was happening in Europe in the same period. Many of the same technical innovations the Americans were exploring were also being used by European composers, but in Europe they took on a different spiritual meaning. We have an interesting exchange of letters between Cage and Pierre Boulez in the 1940s. In this correspondance they're both very excited, there's clearly a shared sensation that they were on the same track. Many of the letters are very technical, describing in detail complicated arithmetic schemes, ideas we would recognize now as algorithmic music, schemes just crying out for a computer to implement them. There was eventually a drifting apart of the two composers, however, because there was a radical disjunction in purpose. The Europeans—Boulez, Stockhausen, Varese, Xenakis—took the new compositional innovations available, using randomness, arithmetic and geometric techniques, and other gleanings from science, as a way to continue—to look at it from the American point of view—the romantic project of self-expression through music. These new ways of working were mere techniques, to be added to the toolkit of a skillful composer who was still in the business of creating masterpieces. But for Cage these techniques had a radically different meaning: they were tools for building complex systems that were as free as possible from human influence, a way of opening our ears to hearing things that have absolutely not been made, or pre-chosen, by human beings.

Cage and other composers working at mid-century gave great import to randomness, and spoke often of the meaning and use of randomness in composition. But I would contend that we almost need to mistrust their own statements about what they were doing. What was really happening was an interest in exploring the nature of complex systems, but the terminology of chaos and complexity theory was not available to these artists at this time. Cage pieces that involved random decision making always used randomness as a way to feed systems akin to Monte Carlo simulations. The most important feature of these systems is not their incorporation of randomness: randomness is merely the food that the pattern-making machinery of the algorithmic composition uses to create the authentically new pattern it creates.

A good example of a system of this sort is the Cage composition which is usually talked of as being the *ne plus ultra* of randomness, *Variations IV*. In this piece the score consists of eight sheets of clear acetate, four of which have one line on them, and four of which have one dot. The performer is asked to prepare a performance score for himself, before the performance, by using these materials. First one chooses the set of performance parameters one wants to subject to the process: they may be pitch, loudness, density, some kind of harmonic measures, anything the performer chooses. These become associated with the lines. Then one throws this stack on a table, and takes measurements of the distance between the dots and the lines, each dot representing a particular musical event. The process continues until enough material is generated to specify the performance. Every performance of *Variations IV*, as you might imagine, is radically different, depending on the instruments used—clarinets, sinewave oscillators, automobile horns, shortwave radios—and the parameters chosen.

Now this piece is random, in some sense, but what's interesting about it, upon reflection, is that it's actually a quite constrained system, determined by the geometry, the possible relations of the points and lines. There ends up being correlations between the presumably random parameters that are based on the constraints of this geometry. The constraints are not analytically understood, certainly not by Cage, perhaps by no-one—I don't know if anyone has ever analyzed this—but what results is a complex system of relationships, which is essentially removed from human taste. And the underlying aim, again, is to open the ears to hearing a new experience.

There are many descendents of Cage in this experimental tradition; one important movement following Cage was minimalism. The idea of American minimalist composition is to pare down the systems involved to the point that all the complexity very clearly arises out of the simple physics of the situation. One example would be the piece *Pedulum Music* (1969) by Steve Reich, in which he hung microphones on long cords from a high ceiling, forming pendula of different lengths. The microphones were set swinging and under each microphone was a loudspeaker which output that microphone's signal, forming a squealing feedback loop whenever the mic passed near the speaker. The piece was over when they all stopped, forming a godawful static howl.

Perhaps the most purely minimalist piece of music in this vein is *Music on a Long Thin Wire* (1979) by Alvin Lucier. I quote from the composer's notes accompanying the double LP recording of this music:

> An 80 foot long wire is driven to oscillation by passing a pure sine wave signal through it while a large fixed horeshoe magnet is mounted nearby. That is an electrical sine wave, not a sound; an electrical circuit is formed through the 80 foot long wire. The electrical oscillation in the fixed magnetic field of the horseshoe magnet induces motion of the wire, which creates the sound.
> A single oscillator tuning was chosen. No alteration of the tuning, or manipulation of the wire or fixed magnet was made in any way. The wire played itself: all changes in volume, timbre, harmonic structure, rhythmic and cyclic patterning, and other sonic phenomena were brought about solely by the modes of vibration of the system.

It's difficult to imagine a more passive notion of composition. Lucier doesn't control anything about the process after it is set in motion. The consequences, and the musical interest, are purely the result of physical law and the contingencies of the moment: the wind, the temperature, the imperfections of the string.

Its interesting to look back now at the progression of this tradition, and what's happened as we've gone along through the century. We start, in Ives, with a musical reinterpretation of a musical/acoustic phenomenon. The representation of the phenomenon is still largely filtered through the composer's musical sensibility. In as much as it partakes of what we're calling our experimental music intent, of hearing something that's beyond human composition, it's done through the medium of the composer's artistic recreation.

In Cage we have a further distancing of the artist from the work, where the composer comes up with an algorithm, building a virtual machine that generates music. The framework of the piece is foreordained, but the relationships that arise between the different parts of the music generated may not be forseen when the scheme was developed. The music is more raw and direct in a way, there's more of a "hands-off" attitude: the shaping work of the composer is more restrained, the raw phenomenon of the piece shines through more directly.

And with Lucier, we have an even more hands-off and physically direct situation, where the body of the music is **only** the physics of an actual performance situation. There is no representation going on at all, there is no reinterpretation. We've really reached the point where the whole interest is in listening to the natural phenomena, and Lucier has reduced the machine or mechanism to put us in that situation to the absolute minimum point. The minimalism of the means forms a nice bit of theatre here, as well: the simplicity of the system involved is transparent, and it becomes absolutely clear that we're letting the physical phenomena speak for themselves.

Looking further back to the influences that led up to the American experimental tradition, one can see the sources for our interest in musically embodying the natural. French music has been a key influence, not only *musique concrete*, early experiments in composition with the tape recorder, but the French impressionism of Satie, Ravel, and Debussy. This French music is concerned with painting sonic pictures, with representing sonic environments as living interacting entities. These composers in turn have their origin in what I would call pre-impressionism, the eighteenth century music of Lully and Rameau. According to James H. Johnson's fascinating history of French musical culture *Listening In Paris*, the critical criterion of success for a musical composition in this period was whether the composition successfully portrayed a particular sonic environment. The eighteenth century conventional mappings of natural sounds into musical expression are obscure by our standards—we, having been trained by recording technologies, have grown quite literal minded about sonic imagery—but the intent has similarities to that of our modern experimental work.

We can perhaps see here a thread that provides some continuity through the stormy nineteenth century period of self-obsession. There is a long tradition of thinking of music as providing an image of real world dynamics, but over time the language and means of representation have changed, and become ever more direct and unmediated. Lucier, is, in a sense, the culmination of this process: in his work the notion of music as a way for us to re-experience natural phenomena is taken completely literally.

The next composer I want to discuss is David Tudor. He was a collaborator with Cage, and a virtuoso pianist and performer of contemporary music, who also developed a performance/composition career of his own doing electronic music. His practice was to string together cheap electronic components into complex and ill-understood circuits which exhibited complex and ill-understood behavior. These networks can be thought of as simulations of some kind: simulations perhaps of things that never existed, if that makes any sense. The dynamic behavior of these complex systems is very explicitly what this music is about.

Tudor has this to say about how his piece *Untitled* (1972) was created:

> The generation of Untitled begins with two chains of components, each chain linked together with multiple feedback loops having variable gain and variable phase-shift characteristics. The configurations of devices and their inter-connections, was conceived of as a "giant oscillator," with ran-

> dom characteristics variable by the performers response and consequent actions.
> The components used, mostly home-brew, were: amplifiers (fixed or variable gain, fixed or variable phase-shift, tuned, saturating types), attenuators, filters (several types), switches, and modulators with variable sideband capability.

Tudor's music is difficult to listen to, consisting usually of extremely distorted, noisy, abrasively electronic sounds. This music is often hated, and I can certainly understand how one could legitimately hate it; it makes few, if any concessions to musical taste, and doesn't attempt to satisfy any conventional musical expectations. This inhuman "otherness" is in some sense the point of the music—it's specifically **not** crafted for your delectation. This music has not been shaped to be an ideal object of contemplation. Traditionally music is a kind of sound tailored to our ears; but this music requests that the audience make the effort to attend to the behavior of a system which is indifferent to its effect on human beings.

Tudor's work represents an attempt to map a dynamical phenomenon—the gyrations of electronic circuits—into a sonic form. As such, it's not necessarily a good fit to our perceptual apparatus. It's the composer's job to make as good a fit as he can, but the misfit, the rawness of the music, the stretching that we have to do as listeners is the spiritual task that this music is about. We are asked to find a way to somehow immerse ourselves in the world created by an alien phenomenon.

To me, this is the core of what is so extraordinary about this tradition: it is calling on us to use our inherent ability to analyze an acoustic scene as a way of getting a view into the workings of a complex system of some kind. It's asking for a new kind of listening, some hybrid of aesthetic attention and natural perception, a way of listening adequate for parsing the sonic traces that make up this music. Musical pieces in this world are not communications from one person (the artist) to another (the listener). They are some strange new kind of object, that is not quite natural and not quite a typical artifact either. In other words, rather than receiving a musical form that was created by another person, we're listening to the hidden structure that arises out of a situation that was certainly initiated by a human composer, but which actually has something of a life of its own.

I think we can see that there is a correspondence between what is going on in the Lucier and Tudor pieces, and some of the practices of contemporary physics. Scientists involved in experimental mathematics, or in doing dynamical systems simulation, are performing a very similar work: they create artificial objects/systems, designed to be contemplated and studied as if they were natural objects.

So our little history has now reached the 1970s, and at this time it was pretty well established, in experimental music circles, that the dynamics of a complex system are interesting in themselves as musical phenomena. The "guerilla electronics" approach of Tudor and his followers, of which I count myself, is one approach, but this is also the time when the analog synthesizer is being developed. Or another way to look at this, which perhaps makes the point clearer, is that at this time analog computers, designed as simulation machines, were being re-purposed for use as musical instruments.

This is the context in which I started making music, and I accepted most of the points I've been trying to make above as givens. In the context of the late 1970s and early 1980s, as I said at the beginning, there was a very exciting musical scene happening, of people playing in different collaborational contexts, often hooking different analog synthesizers together to make one big analog synthesizer which would have unpredictable and interesting behavior.

So it was natural when microcomputers became available for us to extend our practice to include little microcomputers in our big synthesizer patches. The League of Automatic Music Composers, a group I played with that began in the late 70s, was dedicated to just that: we would interconnect little single board microcomputers with music synthesis equipment. It opened up a whole new area for us, introducing us to the possibilities inherent in including these very non-linear devices called microprocessors into our networks.

Hooking up a tangle of ad-hoc connections every time we wanted to rehearse or play a concert started to be a nuisance, especially as more and more people were around who wanted to play music in this way. We really needed a way to connect computers for the purpose of making music. (The commercial standard for this purpose now available, called MIDI, didn't exist at this time.) This is where the Hub came in. The Hub was a name we used for a box built by a small group of us to interconnect separate computer/synthesizer systems; eventually it also became the name of a regular computer network band that made use of it.

The band was a group of six composer/performers who each had our own synthesizers, controlled by our own computers, which were all interconnected through our central "mailbox" computer, the Hub itself. The whole point of this exercise was to build music that arose out of the unpredictable behavior of the interconnected systems. Usually a piece was designed by one person, who came up with a specification for what kind of data could be interchanged between players in a particular piece. The players would then program their own computers to have some behavior that follows that spec—as long as their system followed the spec, which was usually pretty simple, they were free to do whatever they wanted. Often the algorithms in each machine were quite simple, and didn't seem to account for the larger structure that would emerge from the asynchronous communication between the machines.

For example, the piece *Is it Borrowing or Stealing?* (1987) by Hub member Phil Stone, has a very simple design. The Hub was used as a repository for melodic figures and the only requirement was that whenever a player played a melodic figure, he reported to the central Hub what he had played, by putting a copy of the information specifying the figure there. Anyone else could take it, use it, modify it, and play what they want. It's a completely open specification for what each player does: it's just that each player has information about what the other players are doing.

Perhaps I should make clear what I mean by a "player"—I mean a person, and his computer and the program he has written. Usually the process would be almost completely automatic, and the action of each player directed by an algorithm running on a computer. No-one is playing anything on a musical keyboard, but the players—the people—generally have some knobs and switches they use to fine tune the operation of the algorithm running on their system. In a sense we acted more like composers or conductors than performers while in performance, just listening and making fine adjustments from time to time. So the system really includes the people, and the musical reactions of the players would be one element of an overall social/electronic musical network.

The communal aspect was really important to me, and I think it was probably the most interesting aspect of the work with the Hub. It was a social experiment, as much as a technical and aesthetic one, and many of the pieces we did were really about exploring the new social permutations suggested by this new way of music making. I did a piece in the late 80's for the Hub called *The Minister of Pitch* where I looked into apportioning musical responsibility in an unusual way. Players were assigned areas of global responsibility based on different musical parameters: one player was in charge of the pitch relations of all the players, another in charge of everyone's timbral decisions, another in charge of rhythm. Other pieces had game structures, in others players would vote or bid on the musical direction. In this sense the Hub was something of a laboratory for new kinds of collaborative work. It was as if the Hub was one collective instrument, which radically changed its character from piece to piece and demanded different modes of cooperation—sometimes including competition!—between players.

One of the more complex of these "social experiment" pieces was *Hub Renga* (1987). It was based on the Japanese poetry form called renga, in which different people each write one line, each responding to the previous line written by someone else. *Hub Renga* was a live radio performance, in collaboration with the Well, a computer bulletin board and messaging system. The Hub was connected to the Well through a dialup line in the studios of KPFA radio in Berkeley. The public could dial up the Well from their home and type in lines of poetry which would be read aloud on the air; this stream of text also was fed directly into the Hub computers, which were programmed to respond to certain "power words" in the text with musical actions of each Hub composer's choosing. The Well poetry community in the weeks leading up to the performance had actually collectively compiled this list of power words.

What is especially interesting to me about this piece is how it redefines the borders of public and private. We tend to think of communications technology as always giving us more presence with each other, but here is a case where things are a bit different: people are able to act as live performers in a group work in solitude, in their own homes, doing the private act of writing and the public, collective act of performing at the same time.

I've been claiming that the American experimental tradition is purely about natural phenomena, that it's not about self-expression, that it's not about the shaping power of the artist's vision, and so on. But now it's time to admit I've overstated my case to make a point. The fact is that this practice is still an artistic one. It is not science, and these pieces are not mere illustrations of scientific principles; they are attempts to create aesthetic experiences. The emphasis is perhaps on finding a way for us to perceive new and alien structures rather than directly expressing personal musical ideas, but the artistry lies in the balance between the two extremes, between wonder at the unknown and self-expression.

The place a composer chooses on this spectrum is a matter of personal temperament. I greatly admire the work of John Cage, for example, but I also know there was a severity in his method, that had to do with a need on his part to expunge the personal from his work, to distance himself from his own taste and his own emotional landscape. He needed to be carried outside himself. In my own work, I don't want to lose Cage's insight of bringing in the foreign and unexpected, but I also need to engage my own faculties in shaping this material.

Much of my recent work has focused on building what are essentially software musical instruments, that are used in live improvisational situations, usually with ensembles of acoustic instruments. These new computer instruments have their own unpredictable complex behaviors, that are partially under my control and partially following their own nature. Playing them, even when playing by oneself, has something of the quality of conversation with another person, or playing music with another person. As in a conversation, each participant doesn't know what his partner will say next, and therefore doesn't even know what he himself will say next in response. One is always responding to what actually happens, which is not always what one expects.

The French philosopher Jean Baudrillard has said, "It is the fate of our technologies to render the world more illusory." Certainly that's the prevalent trend here at century's end, as we live ever more mediated lives, lives in which more and more of our experience is run through conceptual and electronic filters of various kinds. I'm afraid he may be correct—but in looking at the aesthetic position I've been celebrating for the last hour, I see a hopeful alternate vision. This aesthetic repurposes technology away from mediation and towards a means to perceive the dynamics of the world. By engaging in the creation of aesthetic objects beyond our understanding and control, and then applying our perceptual abilities to the task of understanding them, we are closing a circle of connection with the natural world in a new way.

Art and Complexity
J. Casti and A. Karlqvist (editors)

Modeling Complexity for Interactive Art Works on the Internet

Christa Sommerer and Laurent Mignonneau

Based on the idea that interaction and communication between entities of a system are the driving forces behind the emergence of higher and more complex structures in life, we propose to apply principles of complex system theory to the creation of interactive, computer-generated and audience-participatory artworks and to test whether complexity within an artificial computer-generated system can emerge.

1 INTRODUCTION

Creating virtual life on computers ultimately brings up the question of how life has emerged on earth and how it could have developed from simpler units or particles into increasingly complex structures or whole systems of structures that seem to follow a certain inner rule of organization. This is also the central question in the new complex system sciences. The first part of this chapter analyzes some of the theories and principles of complex systems and then proposes an application of principles of complex systems to the creation of interactive, computer-generated, and audience-participatory artworks on the Internet.

2 COMPLEX SYSTEM SCIENCES

Complex system sciences, as a field of research, has emerged in the past decade. It studies how parts of a system give rise to the collective behaviors of the system and how the system interacts with its environment. Social systems formed (in part) out of people, the brain formed out of neurons, molecules formed out of atoms, the weather formed out of air currents are all examples of complex systems. The field of complex systems cuts across all traditional disciplines of science as well as engineering, management, and medicine. It focuses on certain questions about parts, wholes, and relationships. These questions are relevant to all traditional fields. There are three interrelated approaches to the modern study of complex systems: (1) how interactions give rise to patterns of behavior, (2) understanding the ways of describing complex systems, and (3) the process of formation of complex systems through pattern formation and evolution.[1]

3 DEFINING COMPLEXITY

Although there is no exact definition of what a complex system is, there is now an understanding that, when a set of evolving autonomous particles or agents interact, the resulting global system displays emergent collective properties, evolution and critical behavior that have universal characteristics. These agents or particles may be complex molecules, cells, living organisms, animal groups, human societies, industrial firms, competing technologies, etc. All of them are aggregates of matter, energy, and information that display the following characteristics. They:

- couple to each other;
- learn, adapt, and organize;
- mutate and evolve;
- expand their diversity;
- react to their neighbors and to external control;
- explore their options;
- replicate; and
- organize a hierarchy of higher-order structures.

To find a common principle behind the organizational forces in natural system is a complex task, and it seems as if there are as many theories as there are theorists. Some of the numerous theories on complex system shall be briefly overviewed here. Valuable information on the various approaches and definitions of complex system theory are taken from Edmonds [16].

3.1 ALGORITHMIC INFORMATION COMPLEXITY—THE KOLMOGOROV-CHAITIN-SOLOMONOFF DEFINITION

The best known definition of complexity is the Kolmogorov-Chaitin-Solomonoff (KCS) definition [13, 35, 57] describing algorithmic information complexity (AIC), which places complexity somewhere between order and randomness; that is, complexity increases as Pmin (the shortest algorithm that can generate a digit sequence, S) increases to the length equal to the sequence to be computed; when the algorithm reaches this incompressibility limit the sequence is defined as random. The KCS definition distinguishes between "highly ordered" and "highly complex" structures. In highly ordered structures, Pmin *ll* S.

[1]New England Complex System Institute website: ⟨http://necsi.org/guide/whatis.html⟩.

The AIC of a string of symbols is the length of the shortest program to produce it as an output. The program is usually taken as running on a Turing machine. AIC has been one of the most influential complexity measures (along with computational complexity) and has inspired many variations and enhancements including "sophistication" and "logical depth." Although Solomonoff [57] considered it a candidate among equally supported scientific theories, Kolmogorov [35] and Chaitin [13] considered it a measure of information. It has many interesting formal properties [13], including:

1. The more ordered the string, the shorter the program and hence less complex.
2. Incompressible strings (those whose programs are not shorter than themselves) are indistinguishable from random strings.
3. Most long strings are incompressible.
4. In a range of formal systems you can't prove (within that system) that there are strings above a certain fixed level of complexity (derived basically from the AIC of its axioms).
5. In general it is uncomputable.

AIC has been applied in many ways: to define randomness in a non-probabilistic way [40]; to capture descriptive complexity [38]; to motivate, in Rissanen's statistical version, a principled trade-off between the size of model and its error [50]; to clarify biological complexity [27, 44, 56]; and to use in cognitive models [12], economic models [15], and data compression [69]. Smith [56] argues that the implication that the more complex a structure is, the closer it is to being random, is difficult for biologists to accept. The biologist, moreover, needs to know what the sequence of digits specifies. Papentin [44] points out that for the purposes of comparing biological complexities, the KCS algorithm need only generate a description of the entity in some agreed language, L, rather than generate the entity itself.

3.2 HINEGARDNER AND ENGELBERG'S NUMBER OF PARTS DEFINITION

Perhaps the simplest measure of complexity is that suggested by Hinegardner and Engelberg [27]: the number of different parts. This, of course, depends on what we recognize as parts. Hinegardner and Engelberg suggest that at root, organisms are composed of molecules. They do not concern themselves with differences in the complexity of the molecules. It is significant to note that the brain has a greater variety of proteins than any other organ in the body. It has been estimated that at least 125,000 mRNA transcripts are expressed at different times and in different cells of the brain: up to five times as many as in any other tissue. Furthermore, many of these transcripts are further modified before they are translated, so that several different proteins and/or polypeptides may result from one mRNA tape. Taken together with the great variety of other molecular species in the brain, Hinegardner and Engelberg's simple measure may indeed provide a useful first approximation to complexity. But something is left out: the overwhelming connexity. Hinegardner and Engelberg's measure reminds us of the "exploded" diagrams of pieces of machinery. They give some indication of complexity, but leave out what is perhaps most important: "organization" and "levels of organization" [56].

3.3 TOPOLOGICAL COMPLEXITY

The topological complexity described by Crutchfield [14], is a measure of the size of the minimal computational model (typically a finite automaton of some variety) in the minimal formal language in which it has a finite model. Thus the complexity of the model is both "objectivized" by considering only minimal models but also related to the fixed hierarchy of formal languages. This has a number of disadvantages. First, this does not give a unique complexity for any pattern, as there is not necessarily such a "minimal" formal language. Second, in some formal languages the minimal model is uncomputable. Third, in stochastic languages the minimal model will frequently be a completely random one, so one is forced to trade specificity for complexity to get a reasonable result. Crutchfield also defines a measure of specificity similar to effective measure complexity (EMC), as complementary to the topological complexity. In each case the desire to attribute complexity purely objectively to a physical process seems to force a relativization to either some framework for which privilege is claimed (e.g., a Turing machine), to some aspect of the problem (e.g., granularity of representation) or to consideration given only to the minimal size. This, of course, does not completely eliminate the inherent subjective effects in the process of modeling (principally the language of modeling), and obscures the interplay of complexity, specificity, and error involved.

3.4 COMPUTATIONAL COMPLEXITY

Computational complexity is now a much studied area with many formal results [17, 43, 65]. The foundation of complexity theory is the research into computability theory undertaken from the 1930s onward by Alan Turing, Alonzo Church, and Stephen Kleene, among others. The primary considerations then were the formalization of the notion of a computer (e.g., the Turing machine, Church's lambda calculus) and whether such computers could solve any mathematical problem. Of course, the outcome was that there are problems (of varying degrees of unsolvability) that cannot be solved by a computer (and so perished David Hilbert's program, formulated at the beginning of this century, that aimed to show that all mathematical questions could, in principal, be answered in a mechanical way). A whole host of unsolvable problems have since been presented, most of which hail from pure mathematics and computer science,[2] though more recently some have been arisen in theoretical physics and biology [27].

One aim of computational complexity theory is to classify (solvable) problems according to their intrinsic computational difficulty; that is, questions such as "Given a problem, how much computing power and/or resources do we need to solve it?" are posed. Of course, researchers had always been striving for better (with respect to some appropriate criteria) algorithms for the solution to particular problems, but it was not until the late 1960s and early 1970s that a theory of

[2] ⟨http://www.mcs.le.ac.uk/~istewart/moreIAS/BriefCompTheory.html⟩

complexity was established, the pioneers being Cook [12] and Karp [32]. The seminal work of Cook and Karp led to the consideration of complexity classes and how these classes are related by inclusion (structural complexity theory). Many of the fundamental questions in structural complexity theory remain unanswered (e.g., "Is P strictly contained in NP?"), but progress has been made as new techniques and methods are discovered and developed [7, 8].

3.5 DESCRIPTIVE COMPLEXITY THEORY

Descriptive complexity theory is but one aspect of finite model theory, which is the model theory of finite structures. Model theory is the branch of mathematical logic dealing with the relationship between a formal language and its interpretation in mathematical structures. The history of model theory dates back to the nineteenth century when Bolyai and Lobachevsky developed non-Euclidean geometry and Riemann constructed a model in which the parallel postulate is false but all the other axioms are true. Consequently, the mathematical world was forced to admit that a theory may have more than one model. Later in the nineteenth century, Frege formally developed predicate logic and Cantor did likewise with intuitive set theory. Early developments in model theory included work by Loewenheim, Skolem, Tarski, Gödel, and Malcev (1915–1936). Today the literature is reasonably extensive and the theory of models has been extensively applied in other fields such as set theory, algebra and analysis, and is finding newer applications in computer science [?].

Where model theory studies various logics in the class of all structures and definability in fixed infinite structures, finite model theory studies uniform definability in classes of finite structures. The model theory of finite structures (such as finite graphs or finite groups) is very underdeveloped. Early results (that is, pre-1974) include Trakhtenbrot's theorem (1950) [63]: "The set of first-order sentences, over some signature including a relation symbol that is not unary, which are valid over finite structures is not r.e. but is co-r.e.." These early results appeared sporadically and tended to be "finite considerations" of analogous results in model theory. This is true of Trakhtenbrot's result where the analogous result in model theory is due to Gödel (1930): "The set of valid first-order sentences is r.e. but not co-r.e.." In fact, one of the reasons why finite model theory was not actively pursued in the early days was because essential results of model theory, such as the Completeness and the Compactness Theorems, no longer apply to finite structures. Another reason was that logic was originally developed to provide a solid foundation to mathematics which includes all structures, finite and infinite. Under this philosophy, there was no reason to restrict attention to finite structures.

In 1969, Fagin [18] decided to study spectra (a spectrum of a first-order sentence is the set of cardinalities of it's finite models) and Asser's problem (1955): "Is the class of spectra closed under complementation?" In 1970, his investigations expanded to generalized spectra (i.e., existential second-order spectra where not all relation symbols are quantified out). Probably Fagin's most important result was his characterization of NP as the class of generalized spectra (1974). While Fagin's work is now regarded as seminal, finite model theory remained in the doldrums for some years afterward until researchers such as Immerman [30], Grandjean [20], Ajtai and Gurevich [2], and Makowsky [39] took up the challenge in the early 1980s. Interest in the subject has now exploded, mainly due to the intimate relationship (first hinted at by Fagin) between finite model theory and complexity theory [9]. In fact, there is an established subject area within finite model theory dealing explicitly with this relationship, and this is descriptive complexity theory.

3.6 SHANNON ENTROPY

Shannon Entropy [55] can be seen as the difficulty of guessing a message passing down a channel given the range of possible messages. The idea is that the more difficult it is to guess, the more information a message gives you. This was not intended as a measure of complexity, but has been used as such by subsequent authors. Although Shannon did not envisage his measure of information being used to quantify complexity, some of his successors have either used it as such or based complexity measures on it. The Shannon measure of information is a statistical measure based on the probability of receiving a message. If $p(m1), p(m2), \ldots$ are the probabilities of receiving the messages $m1, m2, \ldots$ then the information carried by the message $n1, n2, \ldots$ is defined as $-\sum \log 2(p(ni))$. The more improbable the message, the more information it gives the recipient.

3.7 GOODMAN'S COMPLEXITY

Goodman [19] has devised an elaborate categorization of extra-logical predicates based on expressiveness. For example, a general predicate is deemed more complex than a symmetric one, as it includes the later as a specific example. Likewise a three place predicate is more complex than a two place one. Goodman builds upon this starting point. The idea is that when faced with two theories that have equal supporting experimental evidence, one should choose the simpler one using this measure. The complexity of a complex statement is merely the sum of the complexities of its component predicates, regardless of the structure of the statement. It is similar in spirit to Kemeny's measure (section 3.8.). A recent defense and reformulation of this idea was made by Richmond [49].

3.8 KEMENY'S COMPLEXITY

In the field of "simplicity," Kemeny [34] attributes an integral measure of complexity to types of extra-logical predicates. He does it on the basis on the logarithm of the number of non-isomorphic finite models that a predicate type has. On the basis of this he gives extra-logical predicates a complexity that could be used to decide between equally supported theories. This is similar in style and direction to Goodman's measure in section 3.7.

3.9 HORN COMPLEXITY AND NETWORK COMPLEXITY

The Horn complexity of a propositional function is the minimum length of a Horn formula (in its working variables) that defines that function. This was defined by Aanderaa and Börger [1] as a measure of the logical complexity of Boolean functions. It is polynomially related to network or circuit complexity, which is the minimum number of logical gates needed to implement a logical function [52]. This is very difficult to compute in most cases, but some upper and lower limits can be proved. This measure depends on the choice of logic gates that you can use to build the circuits from. This measure has an immediate importance for electronic engineers who seek to minimize the expense of logic gates as in Lazarev [37].

3.10 EFFECTIVE MEASURE COMPLEXITY

Grassberger [22] defines the effective measure complexity (EMC) of a pattern as the asymptotic behavior of the amount of information required to predict the next symbol to the level of granularity. This captures an aspect of the scaling behavior of the information required for successful prediction by a Markov process model. A similar approach is taken by Badii and Politti [5]. EMC can be seen as the difficulty of predicting the future values of a stationary series, as measured by the size of regular expression of the required model.

3.11 NUMBER OF INEQUIVALENT DESCRIPTIONS

If a system can be modeled in many different and irreconcilable ways, then we will always have to settle for an incomplete model of that system. In such circumstances, the system may well exhibit behavior that would only be predicted by another model. Thus such systems are, in a fundamental way, irreducible. Accordingly, the presence of multiple inequivalent models were considered by Rosen [51] and Pattee [45] as the key characteristic of complexity. Casti [10] extends this approach and defines complexity as the number of nonequivalent descriptions that an observer can generate for a system it interacts with. The observer must choose a family of descriptions of the system and an equivalence relation on them—the complexity is then the number of equivalence classes the family breaks down into, given the equivalence relation.

3.12 OTHER DESCRIPTIONS OF COMPLEXITY

It would exceed the purpose of this chapter to mention all of the other existing definitions and descriptions of complexity, but some shall be mentioned here briefly: Taxonomical Complexity by Tyler Bonner [64], Operational Complexity by Popper [48], Organizational Complexity,[3] Hierarchical Complexity,[4] Schroedinger's definition that "...neuroanatomy is complexity built on complexity" [54] and the Complexity Mailing List,[5] which discusses on-line how to define complexity and its properties.

[3] ⟨http://www.entek.chalmers.se/~olbj/vsm.htm⟩.
[4] ⟨http://www.ccs.fau.edu/kelso/kelsobook.html⟩.
[5] ⟨http://necsi.org/discuss/discuss.html⟩.

4 PROPERTIES OF COMPLEX SYSTEMS

Instrinsically linked to defining complexity is the search for properties of complex systems. Various scholars have undertaken the task to define these properties, and some of them shall be mentioned here briefly. Again, as for the definitions of complexity (section 3), there is no commonly agreed upon "list" of properties that are thought to completely describe all properties of complex systems.

4.1 VARIETY

A complex system is likely to exhibit a greater variety in terms of its behavior and properties. Thus variety is an indication of complexity (though not always, as sometimes a very complex system is necessary to maintain equilibrium). Variety can be measured by the simple counting of types, the spread of numerical values, or the simple presence of sudden changes. In this way it overlaps with information (section 3.1) and entropic (section 3.6) measures. Applications include: stability of ecosystems [46]; competing behaviors and control [47]; tree structures [28]; number of inequivalent models [11]; the interaction of connectivity and complexity [25]; and evolution [41].

4.2 DEPENDENCY

Heylighen [26] suggests that complexity increases when the variety (distinction), and dependency (connection) of parts or aspects increase, and this in several dimensions. These include at least the ordinary three spatial dimensions, geometrical structure, the dimension of spatial scale, the dimension of time or dynamics, and the dimension of temporal or dynamical scale. In order to show that complexity has increased overall, it suffices to show that—all things being equal—variety and/or connection have increased in at least one dimension. The process of increase in variety may be called differentiation, the process of increase in the number or strength of connections may be called integration. We will now show that evolution automatically produces differentiation and integration, and this at least along the dimensions of space, spatial scale, time, and temporal scale. The complexity produced by differentiation and integration in the spatial dimension may be called "structural," in the temporal dimension "functional," in the spatial scale dimension "structural hierarchical," and in the temporal scale dimension "functional hierarchical."

It may still be objected that distinction and connection are in general not given, objective properties. Variety and constraint will depend on what is distinguished by the observer, and in realistically complex systems determining what to distinguish is a far from trivial matter. What the observer does is to pick up those distinctions that are somehow the most important, creating high-level classes of similar phenomena, and neglecting the differences that exist between the members of those classes. Depending on which distinctions the observer makes, he or she may see their variety and dependency (and thus the complexity of the model) to be larger or smaller, and this will also determine whether the complexity is seen to increase or decrease.

4.3 IRREDUCIBILITY

Irreducibility is a source of complexity. A classic example is the three-body problem in Newtonian mechanics, where the goal is to solve the equations of motion of three bodies that travel under mutual gravitational attraction. This is analytically unsolvable and hence is qualitatively different from any reduction to several separate two-body problems [4]. Nelson [42] argues that irreducibility is a key factor in complex systems and similar approaches include the writings by Anderson [3], who points out the importance of size to qualitative behavior, and Wimsatt [67], who argues that the evolution of multiple and overlapping functions will limit reduction in biology. While Haken [23] sees irreducibility as the result of self-organization and Kampis [31] as the result of the incompatibility of information and computation, Pattee [45] sees it as the result of the epistemic cut between syntax and semantics. Other similar holistic approaches include Rosen's writings [51] who also relates irreducibility to complexity.

4.4 ABILITY TO SURPRISE

It is difficult to model complex systems, so it is likely that any model we have is incomplete in some respect. If we have come to rely on this model (for instance when the system has conformed to the model for some time or under a variety of circumstances) and the system then deviates from that model, we are surprised. The ability to surprise is not possessed by very simple and thus well-understood systems, and consequently comes to be seen as an essential property of complex systems [16].

4.5 SYMMETRY BREAKING

Heylighen [26] argues that complexity can then be characterized by lack of symmetry or "symmetry breaking," by the fact that no part or aspect of a complex entity can provide sufficient information to actually or statistically predict the properties of the others parts. This again connects to the difficulty of modeling associated with complex systems.

4.6 COMPLEXITY AS RELATIVE TO THE FRAME OF REFERENCE

Edmonds [16] notes that complexity necessarily depends on the language used to model a system. He argues that effective complexity depends on the framework chosen from which to view/model the system of study. The criticality of scale in the modeling of phenomena also leads Badii and Politi [5] to focus their characterization of complexity solely on such hierarchical and scaling effects. "The study of the scaling behavior of physical observables from finite-resolution measurements appears, therefore, as an essential instrument for the characterization of complexity."

4.7 MIDPOINT BETWEEN ORDER AND DISORDER

Complexity is sometimes posited as a mid-point between order and disorder. Edmonds [16] notes that the definition of complexity as midpoint between order and disorder depends on the level of representation: what seems complex in one representation may seem ordered or disordered in a representation at a different scale. For example, a pattern of cracks in dried mud may seem very complex. When we zoom out and look at the mud plain as a whole, though, we may see just a flat, homogeneous surface. When we zoom in and look at the different clay particles forming the mud, we see a completely disordered array. Havel [24] states that the paradox can be elucidated by noting that scale is just another dimension characterizing space or time, and that invariance under geometrical transformations, like rotations or translations, can be similarly extended to scale transformations (homotheties).

4.8 COMPLEXITY THROUGH PHASE TRANSITION

According to Kauffman [33], the pure evolutionary view of nature in the Darwinian sense fails to explain the vast structures of order found in nature. By stressing only natural selection, patterns of spontaneous order cannot be sufficiently described or predicted. In Kauffman's view, this order arises naturally as an "order for free." As a consequence, life is an expected phenomenon deeply rooted in the possibilities of the structures themselves. Kauffman argues that, considering how unlikely it is for life to have occurred by chance, there must be a simpler and more probable underlying principle. He hypothesizes that life actually is a natural property of complex chemical systems and that if the number of different kinds of molecules in a chemical soup passes a certain threshold, a self-sustaining network of reactions—an autocatalytic metabolism—will suddenly appear. It is thus the interaction between these molecules that enables the system to become more complex than its mere components taken by themselves.

Kauffman and other researchers at the Santa Fe Institute for Complex Systems Research call the transition between the areas of simple activity patterns and complex activity patterns a phase transition. Kauffman [33] has modeled a hypothetical circuitry of molecules that can switch each other on or off to catalyze or inhibit one of their production. As a consequence of this collective and interconnected catalysis or closure, more complex molecules are catalyzed, which again function as catalyzers for even more complex molecules. Kauffman argues that, given that a critical molecular diversity of molecules has appeared, life can occur as catalytic closure itself crystallizes. A model built by Kauffman is the Boolean network model, which basically describes the connections and relations between three elements. The networks described by Kauffman in the Boolean network model show stability, homeostatis, and the ability to cope with minor modifications when mutated; they are stable as well as flexible. The poised state between stability and flexibility is commonly referred to as the "edge of chaos."

4.9 LIFE AT THE EDGE OF CHAOS

Two of the first scientists to describe the idea of complex patterns and the ones who defined the term "life at the edge of chaos" were Langton [36] and Packard. They discovered that in a simulation of cellular automata there exists a transition region that separates the domains of chaos and order. Cellular automata were invented in the 1950s by von Neumann [66]. They form a complex dynamical system of squares or cells that can change their inner states from black to white according to the general rules of the system and the states of the neighboring cells. When Langton and Packard observed the behavior of cellular automata, they found that although the cellular automata obey simple rules of interaction of the type described by Wolfram [68], they can develop complex patterns of activity. As these complex dynamic patterns develop and roam across the entire system, global structures emerge from local activity rules, which is a typical feature of complex systems. Langton and Packard's automata indeed show some kind of phase transition between three states. Langton and Packard hypothesize that the third stage of high communication is also the best place for adaptation and change and in fact would be the best place to provide maximum opportunities for the system to evolve dynamic strategies of survival. They furthermore suggest that this stage is an attractor for evolving systems. Subsequently, they called the transition phase of this third stage "life at the edge of chaos" [36].

Other researchers at the Santa Fe Institute have extended this idea of life found in this transition phase and applied it to chemistry. In 1992, Fontana developed a logical calculus that can explore the emergence of catalytic closure in networks of polymers [18]. A related approach is seen in the models of physicist Bak [6], who sees a connection between the idea of phase transition, or "life at the edge of chaos," and the physical world, in this case a sand pile onto which sand is added at a constant rate.

5 APPLYING PRINCIPLES OF COMPLEX SYSTEMS TO THE CREATION OF INTERACTIVE AND COMPUTER-GENERATED ARTWORKS

To summarize, we can see that while there are several different definitions and examples of complex systems (Papentin [44], Hinegardner and Engelberg [27], Crutchfiled [14], Papadimitriou [43], Immerman [30], Bonner [64], Grassberger [22], Badii and Politi [5], Heylighen [26], Havel [24], and the comprehensive overview by Edmonds [16]), there is in fact no unified complex systems theory as such. On the other hand, models by Kauffman [33], Langton and Packard [36], Fontana [18], and Bak [6] suggest complex adaptive systems, systems at the "edge of chaos" where internal changes can be described by a power law distribution. These systems are at the point of maximum computational ability, maximum fitness, and maximum evolvability. It is hypothesized that these models could indeed explain the emergence of life and complexity in nature. While, as we have seen, Kauffman's concept of phase transition is not the only model for creating complexity, it does provide an advantageous starting point for creating an artistic system that tries to incorporate some of the features of complex adaptive systems. Based on these considerations we have modeled two artistic interactive systems, which shall be described here in more detail.

6 VERBARIUM—MODELING EMERGENCE OF COMPLEXITY FOR INTERACTIVE ART ON THE INTERNET

Based on the objective to apply principles of complex systems to the creation of interactive artworks on the Internet, we have developed a first prototype to model a complex system for the Internet. Our system, called VERBARIUM [68], is an interactive web site where users can choose to write e-mail messages that are immediately translated into visual three-dimensional shapes. As the on-line users write various messages to the VERBARIUM's web site, these messages are translated by our in-house "text-to-form editor" into various three-dimensional shapes. By accumulation, these shapes can collectively create more complex image structures than the initial input elements. It is anticipated that the users increased interaction with the system will caues increasingly complex image structures to emerge over time.

7 VERBARIUM SYSTEM OVERVIEW

VERBARIUM is available on-line at the following web page: ⟨http://www.fondation.cartier.fr/verbarium.html⟩. The on-line user of VERBARIUM can create three-dimensional shapes in real time by writing a text message within the interactive text input editor in the lower-left window of the web site. Within seconds the server receives this message and translates it into a three-dimensional shape that appears in the upper-left window of the web site. Additionally, this shape is integrated into the upper-right window of the site, where all messages transformed into shapes are stored in a collective image. An example screenshot of the VERBARIUM web site is shown in color plate 4.

VERBARIUM consists of the following elements:

1. A JAVA-based web site (plate 4)
2. An interactive text input editor (lower-left window in plate 4)
3. A graphical display window to display the three-dimensional forms (upper-left window in plate 4)
4. A collective display window to display the collective three-dimensional forms (upper-right window in plate 4)
5. A genetic Text-to-Form editor to translate text characters into design functions

7.1 VERBARIUM'S TEXT-TO-FORM EDITOR

We have set up a system that uses the simplest possible component for a three-dimensional form that can subsequently model and assemble more complex structures. The simplest possible form we constructed is a ring composed of eight vertices. This ring can be extruded in x, y, and z axes, and during the extrusion process the rings' vertices can be modified in x, y, and z axes as well. Through addition and constant modification of the ring parameters, the entire structure can grow, branch, and develop. Different possible manipulations, such as scaling, translating, stretching, rotating, and branching of the ring and segment param-

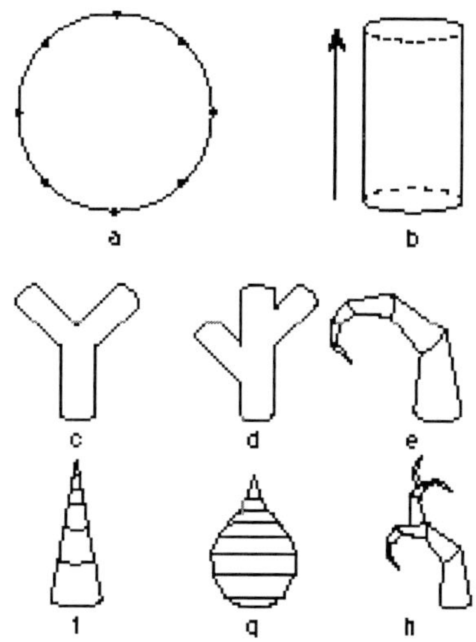

FIGURE 1 Example of VERBARIUMS's growing structures.

eters, creates diverse and constantly growing structures, such as those shown in figure 1.

Figure 1(a) shows the basic ring with eight vertices, and figure 1(b) shows the extruded ring that forms a segment. Figures 1(c) and 2(d) show branching possibilities, with branching taking place on the same place (=internodium) (1(c)) or on different internodiums (1(d)). There can be several branches attached to one internodium. Figure 1(e) shows an example of segment rotation, and figure 1(h) shows the combination of rotation and branching. Figures 1(f) and 1(g) are different examples of scaling. In total, there are about 50 different design functions, which are organized into the design function look-up table (fig. 2). These functions are responsible for "sculpting" the default ring through modifications of its vertex parameters.

The translation of the actual text characters of the user's email message into design function values is done by assigning ASCII values to each text character according to the standard ASCII table shown in figure 3.

Each text character refers to an integer. We can now proceed by assigning this value to a random seed function *rseed.* In our text example from figure 4, *T of This* has the ASCII value 84, hence the assigned random seed function for *T* becomes *rseed(84).* This random seed function now defines an infinite sequence of linearly distributed random numbers with a floating point precision of 4 bytes (float values are between 0.0 and 1.0). These random numbers for the first character of the word *This* will become the actual values for the modification parameters in the design function table. Note that the random number we use is a so-called "pseudo random," generated by an algorithm with 48-bit precision, meaning that if the same *rseed* is called once more, the same sequence of linearly distributed random

function1 translate ring for certain amount (a) in x
function2 translate ring for certain amount (a) in y
function3 translate ring for certain amount (a) in z
function4 rotate ring for certain amount (b) in x
function5 rotate ring for certain amount (b) in y
function6 rotate ring for certain amount (b) in z
function7 scale ring for certain amount (c) in x
function8 scale ring for certain amount (c) in y
function9 scale ring for certain amount (c) in z
function10 copy whole segment(s)
function11 compose a new texture for segment(s)
function12 copy texture of segment(s)
function13 change parameters of RED in segment(s)texture
function14 change parameters of GREEN insegment(s)texture
function15 change parameters of BLUE in segment(s)texture
function16 change patterns of segment(s)texture
function17 exchange positions of segments
function18 add segment vertices
function19 divide segment in x to create branch
function20 divide segment in y to create branch
function21 divide segment in z to create branch
function22 create new internodium(s) for branch(es)
function23 add or replace some of the above functions
function24 randomize the next parameters
function25 copy parts of the previous operation
function26 add the new parameter to previous parameter
function27 ignore the current parameter
function28 ignore the next parameter
function29 replace the previous parameter by new parameter
....
function50

FIGURE 2 VERBARIUM's design function table.

numbers will be called. Which of the design functions in the design function table are actually updated is determined by the following characters of the text, i.e., *his*; we then assign their ASCII values (104 for h, 105 for I, 115 for s ...), which again provide us with random seed functions *rseed(104)*, *rseed(105)*, *rseed(115)*. These random seed functions are then used to update and modify the corresponding design functions in the design function look-up table, between design function1 and function50. For example, by multiplying the first random number of *rseed(104)* by 10, we get the integer that assigns the amount of functions that will be updated. Which of the 50 functions are precisely updated is decided by the following random numbers of *rseed(104)* (as there are 50 different functions available, the following floats are multiplied by 50 to create integers). Figure 4 shows in detail how the

<table>
<tr><td>33 !</td><td>34 "</td><td>35 #</td><td>36 $</td><td>37 %</td><td>38 &</td><td>39 '</td><td></td></tr>
<tr><td>40 (</td><td>41)</td><td>42 *</td><td>43 +</td><td>44 ,</td><td>45 -</td><td>46 .</td><td>47 /</td></tr>
<tr><td>48 0</td><td>49 1</td><td>50 2</td><td>51 3</td><td>52 4</td><td>53 5</td><td>54 6</td><td>55 7</td></tr>
<tr><td>56 8</td><td>57 9</td><td>58 :</td><td>59 ;</td><td>60 <</td><td>61 =</td><td>62 ></td><td>63 ?</td></tr>
<tr><td>64 @</td><td>65 A</td><td>66 B</td><td>67 C</td><td>68 D</td><td>69 E</td><td>70 F</td><td>71 G</td></tr>
<tr><td>72 H</td><td>73 I</td><td>74 J</td><td>75 K</td><td>76 L</td><td>77 M</td><td>78 N</td><td>79 O</td></tr>
<tr><td>80 P</td><td>81 Q</td><td>82 R</td><td>83 S</td><td>84 T</td><td>85 U</td><td>86 V</td><td>87 W</td></tr>
<tr><td>88 X</td><td>89 Y</td><td>90 Z</td><td>91 [</td><td>92 \</td><td>93]</td><td>94 ^</td><td>95 _</td></tr>
<tr><td>96 `</td><td>97 a</td><td>98 b</td><td>99 c</td><td>100 d</td><td>101 e</td><td>102 f</td><td>103 g</td></tr>
<tr><td>104 h</td><td>105 i</td><td>106 j</td><td>107 k</td><td>108 l</td><td>109 m</td><td>110 n</td><td>111 o</td></tr>
<tr><td>112 p</td><td>113 q</td><td>114 r</td><td>115 s</td><td>116 t</td><td>117 u</td><td>118 v</td><td>119 w</td></tr>
<tr><td>120 x</td><td>121 y</td><td>122 z</td><td>123 {</td><td>124 |</td><td>125 }</td><td>126 ~</td><td></td></tr>
</table>

FIGURE 3 ASCII table.

Example word: This

T => rseed(84) => {0.36784, 0.553688, 0.100701,...}
(actual values for the update parameters)

h => rseed(104) => {0.52244, 0.67612, 0.90101,...}
# 0.52244 * 10 => get integer 5 => 5 different
functions are called within design function table

0.67612 * 50 => get integer 33 => function 33
within design function table will be updated by value 0.36784 from 1. value of rseed(84)

0.90101 * 50 => get integer 45 => function 45
within design function table will be updated by value 0.553688 from 2. value rseed(84)
.... until 5. value

FIGURE 4 Example of assignment between random functions and design functions.

entire assignment of random numbers to design functions operates. As mentioned above, the actual float values for the update parameters come from the random seed function of the first character of the word, *rseed(84)*.

As explained earlier, the basic "module" is a ring that can grow and assemble into segments that can then grow and branch to create more complex structures as the incoming text messages modify and "sculpt" the basic module by the design functions available in the design function table in figure 2.

7.2 VERBARIUM'S COMPLEXITY POTENTIAL

Depending on the complexity of the incoming text messages, the three-dimensional forms become increasingly shaped, modulated and varied. As there is usually great variation among the texts, the forms themselves also vary greatly in appearance. As a result, each individual text message creates a very specific three-dimensional structure that can at times look like an organic tree or at other times look more like an abstract form. All forms together build a collective image displayed in the upper-right window of the web site: it is proposed that the complex image structure that emerges represents a new type of structure that is not solely an accumulation of its parts but instead represents the amount and type of interactions of the users with the system. Another example of forms created by a different text message is shown in color plate 5, this time with the text written in French.

7.3 VERBARIUM'S COMPLEXITY EVALUATION

VERBARIUM enables on-line users to create three-dimensional shapes by sending text messages to theVERBARIUM web site. Using our text-to-form editor, this system translates the text parameters into design parameters for the creation and modulation of three-dimensional shapes. These shapes can become increasingly complex as the users interact with the system. A collective image hosts and integrates all of the incoming messages that have been transformed into three-dimensional images, and as users increasingly interact with the system an increasingly complex collective image structure emerges. As it is no longer possible to deconstruct the collective image into its initial parts, some of the features of complex systems, such as variety and dependency (as described in sections 4.1. and 4.2.), as well as irreducibility (as described in section 4.3.) and symmetry breaking (as described in section 4.5.) are thought to have emerged.

8 LIFE SPACIES II—AN INTERACTIVE EVOLUTIONARY ENVIRONMENT ON THE INTERNET

While some features of a complex system have clearly emerged within VERBARIUM, other defining aspects associated with complex system (as described in section 3) have so far been left out. These features include: the ability to couple to each other, to learn, adapt and organize, to mutate and evolve, to expand their diversity, to react to their neighbors and to external control, to explore their options, to replicate, and to organize a hierarchy of higher-order structures.

To deal with these issues and to model a system that includes these aspects, we created a second system called "Life Spacies II." This work was commissioned by the ICC-NTT InterCommunication Museum in Tokyo, and the first version "Life Spacies" was first shown in spring 1997 [69]. The artificial life and genetic programming techniques we used for "Life Spacies II" came from our previous interactive evolutionary systems, such as the ones found in literature [60, 61].

8.1 LIFE SPACIES II—GUI TO CREATE AND FEED CREATURES

"Life Spacies II"[6] consists of a graphical user interface (GUI) that allows users to type text messages into the Internet web page text editor create three-dimensional forms, in this case artificial creatures which start to live and itneract within an artificial environment. As in VERBARIUM, written text is used as genetic code to create three-dimensional forms. In addition to creating creatures, users can feed their creatures by releasing text characters on the GUI. Food particles are in fact

[6] As shown in color plate 6.

text characters, and the user can decide how much text, which type of text, and where to place the text by typing specific text characters within the GUI of the web page. Instantaneously, the text (food) is shown and picked up by the creatures on the large projection screen as shown in color plate 7.

8.2 LIFE SPACIES II—TEXT-TO-FORM EDITOR

While the abstract three-dimensional shapes in VERBARIUM are based on ring structures that develop and deform through user interaction, the default form in Life Spacies II is a sphere. As in VERBARIUM, the "text-to-form editor" is based on the idea of linking the characters and the syntax of a text to specific parameters in the creature's design. In a way similar to the genetic code in nature, letters, syntax and sequencing of text are used to code certain parameters in the creature's design functions. The text parameters and their combinations influence form, shape, color, texture, and the number of bodies and limbs. The default form of a creature is a body composed of a sphere consisting of 100 vertices, that is, 10 rings with 10 vertices each. All vertices can be modified in x, y, and z axes to stretch the sphere and create new body forms. Several bodies can be attached to each other or a pair of limbs can be created. According to the sequencing of the characters within the text, the parameters of x, y, and z for each of the 100 vertices can be stretched and scaled, the color values and texture values for each body and limb can be modified, the number of bodies and limbs can be changed, and new locations for attachment points of bodies and limbs can be created. Since each of the vertex parameters is changeable and all of the bodies and limbs can be changed as well, about 50 different design functions for the creature's design parameters are available. As there is great variation in the texts sent by different people, the creatures themselves also vary greatly in their appearance and behavior. Detailed information on how we encode written text into the genetic code that determines the design functions for the creation of the three-dimensional creatures can be found in literature [62].

8.3 BEHAVIOR OF THE CREATURES

8.3.1 Energy and Speed. A creature's behavior is basically dependent on two parameters: (a) its Energy level (E) and (b) its Speed (S) or ability to move. While the Energy level (E) is a value that constantly changes as the creature moves in its environment and decreases by increased movement, the Speed (S) value is designed by the creature's body physics. A creature with a large body and small limbs will typically move more slowly than a creature with a small body and long limbs. Additionally, the shape of the creature's body and limbs have an influence on its ability to move. On the other hand, the Speed (S) value is set at creation through the arrangement of text characters in the creatures genetic code, which is interpreted and translated by the design function table as explained in literature [5].

8.3.2 Interaction among Creatures. The creature's interaction with other creatures is based on how much Energy (E) it has at a given moment and the Speed (S)

Speed (S): depends on creatures body and limb size decides how fast the creature can move
Energy (E): E = 1 at birth
Speed (S) of movement reduces E
E < 1 ..... creature becomes hungry
E > 1 ..... creature can mate

FIGURE 5 Creatures behavior decision parameters.

Feeding: if E < 1 .... creature wants to eat text characters
it eats the same characters as its genetic code
("John" creature eats: "J', "o", "h", "n")
Mating: E > 1 .... creature wants to mate, if successful,
parents will exchange their genetic code
-> a child creature can be born
Evolution: Selection on faster creatures, as they can eat and mate more frequently

FIGURE 6 Creatures interaction parameters.

it can move in the environment. If, for example, the creatures Energy level (E) reaches a certain value of $E < 1$, the creature becomes hungry and wants to eat. On the other hand, if the Energy level rises to $E > 1$, the creature wants to mate with other creatures. Figures 5 and 6 show this relationship between energy levels and feeding and mating behavior.

8.3.3 Feeding. A creature whose Energy level has risen to $E < 1$ becomes virtually hungry and desires to eat text characters provided by the user through the system's text input editor. The kind of text characters released depends solely on the user's decision. Creatures also have preferences for certain types of food and only eat text characters contained in their genetic message. For example, a creature whose genetic code is "John" will only eat "J," "o," "h," and "n" characters. By eating text characters, the creature will manage to accumulate a certain amount of energy, and eventually its Energy level can again rise to $E > 1$. However, it might be necessary for the creature to eat several text characters as the creatures vigorously move while looking for text characters.

8.3.4 Mating. Given that a creature succeeds in adding energy to the level of $E > 1$, it becomes a potential mating partner. It will now look for a suitable mate, whose energy level is also above 1. The two potential parent creatures will now move toward each other and try to collide. If successful, the two parents exchange their genetic code through a crossover operation and, as a result, a child creature is born. This offspring creatures carries the genetic code of its parents with some mutations. Figure 7 shows an example of a mating process.

8.3.5 Growth and Death. A creature's lifetime is not predetermined but influenced by how much it eats. Through eating the creature increases its body size until

```
Parent creature (P1) and (P2), child (C)
| .... area of cross-over
^ ... location of mutation
Parent Creatures P1 and P2
(P1) This is a creature, it lives in Tokyo.
             ^     |          |              ^
(P2)  This creature is now living in Tokyo.
                   |          |
Child Creature C
(C) This is ancrea is now I lives in Toky.
```

FIGURE 7 Mating process and birth of child creature.

reaching a maximum size of about four times the original body size. On the other hand, a creature will starve when it does not eat enough text characters and ultimately die and sink to the ground.

8.4 LIFE SPACIES II'S COMPLEXITY EVALUATION

The constant movements, feeding, mating, and reproduction of the creatures result in a complex system of interactions that displays features of artificial evolution where selection favors faster creatures. Additionally, the users' input decisions on how to write the text messages and on how to feed the creatures also add constant change to the system. As a result, a system is created that features complex interactions between creatures as well as between users and creatures. Color plate 8 shows a screen shot of different creatures as they mate and feed on text characters. When we go back to the definitions of complex systems given in section 3, we see that "Life Spacies II" displays the following features associated with complex systems: to adapt and organize, to mutate and evolve, to expand their diversity, to react to their neighbors and to external control and to explore their options and to replicate.

9 CONCLUSIONS AND OUTLOOK

We have introduced two interactive systems for the Internet that enable on-line users to create three-dimensional shapes or creatures by sending text messages to the systems' web sites. While various artists and designers have created artificial life artworks or entertainment software since the mid 1990s, most of these software products have provided the users with predefined creatures or parts of creatures, to be chosen or assembled by the users [21, 29].[7][8][9] As this only allows limited design decisions by the users of the systems, we created "VERBARIUM" and "Life Spacies II," more flexible systems that give users more design and interaction decisions. Written text, provided at random by the users of the system, is used as genetic code, and our "text-to-form editor" translates the written texts into three-dimensional autonomous creatures whose bodies, behaviors, interactions, and survival are solely based on their genetic code and the users' interactions. As the users interact with these systems, the systems themselves become increasingly complex, displaying features of complex systems such as variety and dependency, irreducibility, symmetry breaking, adaptation and organization, mutation and evolution, expansion of diversity, reaction to neighbors and to external control, exploration of their options, and replication. Our future research will concentrate on expanding these systems by designing structures that additionally allow for learning and the organization of hierarchies of higher-order structures within these systems.

ACKNOWLEDGMENTS

This research has been conducted at the ATR Media Integration and Communications Research Laboratories in Kyoto, Japan. We would especially like to thank Tom Ray and John Casti for their continuous and valuable discussions.

REFERENCES

[1] Aaderaa, S. O., and E. Bröger. "The Equivalence of Horn and Network Complexity of Boolean Functions." *Acta Informatica* **15** (1981): 303–307.
[2] Ajtai, M., and Y. Gurevich. "Monotone versus Positive." *ACM* **34** (1987): 1004–1015.
[3] Anderson, P. W. "More is Different." *Science* **177** (1972): 393–396.
[4] Badii, R. "Complexity and Unpredictable Scaling of Hierarchical Structures." In *Chaotic Dynamics, Theory and Practice*, edited by T. Bountis, 1–19. New York and London: Plenum Press, 1992.
[5] Badii, R., and A. Politi. *Complexity: Hierarchical Structures and Scaling in Physics.* Cambridge: Cambridge University Press, 1997.
[6] Bak, P., and K. Chen. "Self-Organized Criticality." *Sci. Am.* **January** (1991): 46–54.
[7] Balcazar, J. L. *Structural Complexity*, vol I. Berlin: Springer-Verlag, 1990.
[8] Balcazar, J. L. *Structural Complexity*, vol II. Berlin: Springer-Verlag, 1990.
[9] Banzhaf, W. "Self-Replicating Sequences of Binary Numbers: The Build-up of Complexity." *Complex Systems* **8** (1994): 215–225.

[7] Life Drop home page, ⟨http://www.virtual-worlds.net/lifedrop/⟩.

[8] PostPet email software from Sony Communication Network Corporation, ⟨http://www.sony.com.sg/postpet/⟩.

[9] Darwin Pond by Jeffrey Ventrella, ⟨http://www.ventrella.com/Darwin/darwin.html⟩.

[10] Casti, J. L. "On System Complexity: Identification, Measurement and Management." In *Complexity Language and Life: Mathematical Approaches*, edited by J. Casti and A. Karlquist, 146–173. Berlin: Springer, 1986.

[11] Casti, J. L. "The Simply Complex: Trendy Buzzword or Emerging New Science?" *Bull. SFI* **7** (1992): 10–13.

[12] Cook, S. *The Complexity of Theorem Proving Procedures*, Proc. Third Annual ACM STOC Symposium, 151–158, 1971.

[13] Chaitin, G. J. "On the Length of Programs for Computing Finite Binary Sequences." *J. Assoc. Comp. Mach.* **13** (1966): 547–569.

[14] Crutchfield, J. P. "The Calculi of Emergence: Computation, Dynamics and Induction." *Physica D* **75** (1994): 11–54.

[15] De Vany, A. "Information, Bounded Rationality, and The Complexity of Economic Organisation." Discussion paper, Department of Mathematics, University of California, 1993.

[16] Edmonds, B. "Syntactic Measures of Complexity." Doctoral Thesis, University of Manchester, Manchester, UK, 1999.

[17] Fagin, R. "Contributions to the Model Theory of Finite Structures." Ph.D. Thesis, University of California Berkeley, 1973.

[18] Fontana, W. 1992. "Algorithmic Chemistry." In *Artificial Life II*, edited by C. G. Langton, J. D. Farmer, S. Rasmussen, and C. Taylor, 159–209. Santa Fe Institute Studies in the Sciences of Complexity, Proc. Vol. X. Reading, MA: Addison-Wesely, 1992.

[19] Goodman, N. *The Structure of Appearence.* Indianapolis: Bobbs-Merrill, 1966.

[20] Grandjean, E. "The Spectra of First-Order Sentences and Computational Complexity." *SIAM J. of Comp.* **13(2)** (1984): 356-373.

[21] Grand, S., D. Cliff, and A. Malhotra. "Creatures: Artificial Life Autonomous Software Agents for Home Entertainment." Millennium Technical report 9601, University of Sussex Technical Report CSRP434, 1996.

[22] Grassberger, P. "Towards a Quantitative Theory of Self-Generated Complexity." *Intl. J. Theoret. Phys.* **25** (1986): 907–938.

[23] Haken, H. *Information and Self-Organisation: A Macroscopic Approach to Complex Systems.* Springer Series in Synergetics. Berlin: Springer-Verlag, 1988.

[24] Havel, I. "Scale Dimensions in Nature." *Intl. J. Gen. Sys.* **23(2)** (1995): 303–332.

[25] Heylighen, F. "Relational Closure: A Mathematical Concept for Distinction-making and Complexity Analysis." In *Cybernetics and Systems '90*, edited by R. Trappel, 335–341. Singapore: World Scientific, 1990.

[26] Heylighen, F. "The Growth of Structural and Functional Complexity during Evolution." In *The Evolution of Complexity*, edited by F. Heylighen and D. Aerts. Kluwer Academic Publishers, 1996.

[27] Hinegardner, R., and H. Engelberg. "Biological Complexity." *J. Theoret. Biol.* **104** (1983): 7–20.

[28] Huberman, B. A., and T. Hogg. "Complexity and Adaption." *Physica D* **22** (1986): 376–384.

[29] Hurry, M., J. Prophet, and G. Selley. "TechnoSphere." In *AlifeVII Artificial LifeVII in-house Conference Proceedings.* Portland, Oregon, 2000.

[30] Immerman, N. "Descriptive Complexity." Springer Graduate Texts in Computer Science, 1999.

[31] Kampis, G. "Information, Computation and Complexity." In *Nature, Cognition and Systems*, edited by M. E. Carvallo, 313–320. Dordrecht: Kluwer, 1988.

[32] Karp, R., and R. Lipton. "Turing Machines That Take Advice." *Ensiegn. Math.* **28** (1982): 192–209.

[33] Kauffman, S. *At Home in the Universe. The Search for Laws of Complexity.* New York: Oxford University Press, 1995.

[34] Kemeny, J. G. "Two Measures of Complexity." *J. Phil.* **52** (1953): 722–733.

[35] Kolmogorov, A. N. "Three Approaches to the Quantitative Definition of Information." *Prob. Infor. Trans.* **1** (1965): 1–17.

[36] Langton, C. "Life at the Edge of Chaos." In *Artificial Life II*, edited by C. G. Langton, J. D. Farmer, S. Rasmussen, and C. Taylor, 41–91. Santa Fe Institute Studies in the Sciences of Complexity, Proc. Vol. X. Reading, MA: Addison-Wesely, 1992.

[37] Lazarev, V. I. "Complexity and Synthesis of Minimal Logic Circuits using Multiplexers." *Cybernetics* **28** (1992): 796–799.

[38] Löfgren, L. "Complexity of Descriptions of Systems: A Foundational Study." *Intl. J. Gen. Sys.* **3** (1974): 197–214.

[39] Makowsky, J., and Y. Pnueli. "Arity versus Alternation in Second-Order Logic." In *Logical Foundations of Computer Science*, edited by A. Nerode and Y. Matiyasevich, 240–252. Springer LNCS 813, 1994.

[40] Martin-Lof, P. "The Definition of Random Sequences." *Infor. & Cont.* **9** (1966): 602–619.

[41] Mcshea, D. "Complexity and Evolution: What Everybody Knows." *Biol. & Phil.* **6(3)** (1991): 303–3024.

[42] Nelson, R. J. "Structure of Complex Systems." *Phil. Sci. Assoc.* **2** (1976): 523–542.

[43] Papadimitriou, Ch.H. *Computational Complexity.* Addison Wesley, 1994.

[44] Papentin, F. "On Order and Complexity, General Considerations." *J. Theoret. Biol.* **87** (1980): 421–456.

[45] Pattee, H. "Dynamic and Linguistic Modes of Complex Systems." *Intl. J. Gen. Sys.* **3** (1977): 259–266.

[46] Pimm, S. "The Complexity and Stability of Ecosystems." *Nature* **307** (1984): 321–326.

[47] Poon, L., and C. Grebogi. "Controlling Complexity." *Phys. Rev. Lett.* **75** (1995): 4023–4026.

[48] Popper, K. R. *The Logic of Scientific Discovery.* New York: Basic Books, 1959.

[49] Richmond, S. A. "A Simplification of the Theory of Simplicity." *Synthese* **107** (1966): 373–393.

[50] Rissanen, J. "Complexity of Models." In *Complexity, Entropy and the Physics of Information*, edited by W. H. Zurek, 117–125. Santa Fe Institue Studies in the Sciences of Complexity, Proc. Vol. VIII. Redwood City, CA: Addison-Wesley, 1990.

[51] Rosen, R. "Complexity and System Descriptions." In *Systems: Approaches, theories, Applications*, edited by W.E. Harnett, 169–175. Dordrecht, Holland: D. Reidel, 1977.
[52] Savage, J. E. *The Complexity of Computing.* Krieger Publishing Company, 1987.
[53] Sambrook, T., and A. Whiten. "On the Nature of Complexity in Cognitive and Behavioural Science." *Theor. & Psychol.* **7** (1977): 191–213.
[54] Schroedinger, E. *What is Life?* Cambridge: Cambridge University Press, 1944.
[55] Shannon, C. E., and W. Weaver. *The Mathematical Theory of Communication.* Urbana, Illinois: University of Illinois Press, 1949.
[56] Smith, C. U. M. *Elements of Moqlecular Neurobiology.* Chichester: Wiley, 1990.
[57] Solomonoff, R. J. "A Formal Theory of Inductive Inference." *Infor. & Cont.* **7** (1964): 224–254.
[58] Sommerer, C., and L. Mignonneau. "VERBARIUM and LIFE SPACIES: Creating a Visual Language by Transcoding Text into Form on the Internet." In *1999 IEEE Symposium on Visual Languages*, 90–95. Tokyo, Japan, 1999.
[59] Sommerer, C., and L. Mignonneau. "Life Spacies." In *ICC Concept Book, NTT-ICC Tokyo*, 96–101, 1997. Life Spacies home page: ⟨http://www.ntticc.or.jp/~lifespacies/⟩.
[60] Sommerer, C. and L. Mignonneau. "Interacting with Artificial Life: A-Volve." *Complexity J.* **6(2)** (1997) 13–21.
[61] Sommerer, C., and L. Mignonneau. "The Application of Artificial Life to Interactive Computer Installations." *ALife & Robot. J.* **2(4)** (1998): 2:151–156.
[62] Sommerer, C., and L. Mignonneau. "Life Spacies: A Genetic Text-to-Form Editor on the Internet." In *Proceedings of the AROB 4th Artificial Life and Robotics Conference*, edited by M. Sugisaka, 73–77. Oita, Japan, 1999.
[63] Trakhtenbrot, B. A. *Algorithms and Automatic Computing Machines.* Boston: Heath Series: Topics in Mathematics, 196.
[64] Tyler Bonner J. *The Evolution of Complexity.* Princeton, NJ: Princeton University Press, 1988.
[65] Von Neumann, J. "The General and Logical Theory of Automata." In *The World of Mathematics*, edited by J. R. Newman, Vol. 4, 2070–2098. New York: Simon and Schuster, 1956.
[66] Von Neumann, J. *Theory of Self-Reproducing Automata*, completed and edited by A.W. Burks. Champaign-Urbana: University of Illinois Press, 1966.
[67] Wimsatt, W. "Complexity and Organisation." In *Studies in the Philosophy of Sciences*, edited by K. Schaffner and R. Cohen, 67–86. Dordretch: Reidel, 1972.
[68] Wolfram, S. *Theory and Applications of Cellular Automata.* Singapore: World Scientific Press, 1986.
[69] Zvonkin, A. K., and L. A. Levin. 1970. "The Complexity of Finite Objects and the Development of the Concepts of Information and Randomness by Means of the Theory of Algorithms." *Russian Mathematics Surveys* **256** (1970): 83–124.

Art and Complexity
J. Casti and A. Karlqvist (editors)

Poetic Voice and the Complexity of Bird Song

Marcia Southwick

A Native American legend says that during the first full moon of winter, the chickadee's tongue splits in half, forking like a snake's. During the second full moon, it splits into thirds, only to become whole again in spring. This story about the chickadee's song, though seemingly fanciful, is based on pure observation. As winter approaches, the chickadee's call becomes more elaborate. Then it becomes simpler again as the temperature increases.

How varied can the chickadee's song become without losing its identity? What prevents it from straying so far that it suddenly sound like gibberish to the other chickadees?

Random variations on the chickadee's song must have occurred again and again. Over centuries, the variations on a territorial song that worked to attract a mate or scare away rivals were naturally selected. As accidents occurred, worked well, and promoted survival, they became part of the song, changing it for the better.

But where is the line between a chickadee's song and the song of another species? And how do members of the species recognize each other as belonging to the group? If the chickadee's tongue were suddenly to split into four parts—that is, if its song were to become so complex as to be unrecognizable to other chickadees, the song would lose its identity.

A poem, like the chikadee's song, has restrictions. If a poem ventures too far outside the boundaries of its perceived identity it suddenly seems "other," and its sense of definition collapses. It suddenly crosses over the line between complexity and a formless pile of words. But what do I mean by complexity?

Complexity, as I understand it, is a relatively new branch of theoretical science currently being studied at the Santa Fe Institute, where experts in areas as diverse as physics, microbiology, ornithology, artificial intelligence, economics, linguistics, art, and archaeology are discovering how fundamental "simple" laws in combination with chance events give rise to highly organized forms. The description of how those forms evolve from simple to complex is becoming, in a sense, a new language that creates bridges between areas of study that have not been clearly perceived as connected until now. This new language, in particular, connects science to the arts.

Recently, my husband and I co-chaired a seminar sponsored by the Santa Fe Institute and SITE Santa Fe, a public art space. With the intention of establishing a channel of communication between artists and scientists, we invited painters, architects, musicians, computing experts, biologists, physicists, and others to talk about how principles of simplicity and complexity apply to their particular fields of interest.

An architect discussed the evolution of the design of one of his buildings. A painter traced the steps of his composition process, which started with the assumption that the blank canvas was *something* and not *nothing.* A scientist from the MIT media lab described the future evolution of children's toys and film animation. A neurologist flashed a map of a rat's brain and a human's brain on a screen, and oddly enough, they looked very similar!

The difference between the two brains, it turns out, doesn't necessarily lie solely in the structure of the individual cells, but in the instructions or "rules" that cause those cells to gather into increasingly higher and higher levels of organization and differentiation. (Likewise, rules can express the difference between a random pile of words and a meaningful arrangement of words in the form of poetry.)

The architect also showed us a slide of a "complicated" structure, pointing out its failures. Not only did it have too many random features, but also many of its regularities didn't seem to contribute to any overall sense of purpose. After criticizing certain contemporary architects for designing structures like the one before us, he then informed us that we weren't looking at architecture at all, but at broken lanes of an Los Angeles highway after an earthquake. Clearly, a difference exists between artistic creations that convey significance and those that do not. For significance to occur, regularities must repeat themselves with enough consistency to create meaningful patterns. And yet accidents must also occur to create local regularities that increase complexity.

In the production of a poem, as in the production of a painting, or a musical piece, if too many accidents are assimilated into the process, the piece will be entropic, too random for any "laws" to pull things together. And yet, if the voice of a poem is so law-abiding that it lacks surprises or variations, the writing won't convey passion. Poetic voice is forever hovering along the border between randomness and regularity, searching for just the right tension between them. This balancing act between regularity and randomness produces the mystery and strangeness of voice in poetry, and creates, as a silent counterpart, a distinct individualized *form.*

But what regularities repeat themselves in poems? Contemporary poets establish consistency in many obvious ways—by repeating gestures that establish consistent tone, accent, level of detail, style of imagery, and musical effect. Less obvious are the decisions the poet makes, beginning in the first line. Throughout the poem, fundamental decisions that generate and define the future content are at work. Who is speaking? What is the subject and how is it approached? Is the poem placing the reader into a world that is surreal and dreamlike? Or is its world realistic, one we immediately recognize? Is the poem primarily narrative, lyric, or dramatic?

Is the poem speaking in the language of images. or the language philosophical and abstract? Is the level of diction highly rhetorical? Or is the established level of diction flat and plain? One way to examine the regularities in a poem is to search for particular choices the writer has made, line by line.

It takes a little practice to recognize the regularities operating in contemporary poetry, just as it takes a little practice to distinguish the regularities operating at the base of the variations and branching tongues of the chickadee.

A novice listening to bird calls in a forest probably wouldn't be able to distinguish one bird call from another, much less the variations occurring within a chickadee's song. In the same way, reader of contemporary poetry might be struck at first, in many cases, only by the absence of rhyme and meter.

Ornithologists, like expert practitioners of poetry, can discern subtle distinctions that seem impossible for the lay person to grasp. My first few days of bird watching were excruciating because my experienced friends would all spot and identify a bird before I could raise my binoculars. They seemed to be making things up, arguing about the exact colors of eye-rings, tail-markings, flight characteristics, wing shapes, and so forth, while I was seeing only a little brown blob, wondering what all the fuss was about.

Gradually, after practice, I accumulated "search images." After one has identified a feature, both in poetry and while birding, one can spot that feature much more easily the next time. Now, for example, after a few years of practice, I can spot birds quickly and even have learned to distinguish a few profiles from far away. I can identify several species by the way they fly, and I recognize a song now and then. I still find smaller features difficult to discern (I, therefore, tease my expert birding companions, accusing them of inventing certain eye-rings, beak shapes, and buffy breasts to fit descriptions of birds they *want* to see—the very ones they need for their life lists. But, of course, I know that the smaller features they are seeing, even though undetectable to me, are really *there*.)

A few years ago, a whole new level of ornithological complexity opened up to me when my friends told me about mockingbird song. Apparently the mockingbird can incorporate bits and pieces of other bird songs into its repetoir. It can imitate more than thirty birds, and also will imitate the sounds of saws, crickets, chickens, and frogs. One famous bird in Boston sang on and off for more than five years and imitated at least fifty-one species.

If the mockingbird is mainly an imitator, what makes its song a mockingbird's song and not just an accumulation of stolen phrases? The mockingbird has a certain pattern to its song, even though that pattern is filled with borrowed content. It usually sings phrases in threes or fives, which is why it's so easily recognizable.

Suddenly I'm reminded of poetry—my own in particular, which borrows slogans and phrases from advertising and puts them into unexpected contexts. The poetry of T. S. Eliot has mockingbird tendencies, as he borrowed from so many sources while writing "The Waste Land." Marianne Moore's poetry often consisted of quotes woven together from different sources. And yet the voice of Marianne Moore is recognizeable because she wove the borrowed details together in such a way that the relationships between them became hers and hers alone. The result is that Marianne Moore voice is distinct from anyone else's.

Relationships between the parts in a poem are created just as they are in other aesthetic media. Just as a painter learns to understand the emotional impact of adding a color that changes the relationships of the colors that were placed before it on the canvas, so the poet understands the ripple effect of each poetic line as it alters and builds upon the lines that came before, and the musician understands the changing relationships between repeated phrases as the piece builds, changing the context in which the repeated phrases take on a different resonance. And the ballet dancer understands the variations and connections between the five positions, the basic points from which all of the possible esthetically pleasing dance movements can unfold.

Matisse's description of how he paints would probably strike most artists across the media as a perfect example of how the process of esthetic expression works:

> If, on a clean canvas, I put at intervals patches of blue, green and red, with every touch that I put on, each of those previously laid on loses in importance. Say I have to paint an interior; I see before me a wardrobe. It gives me a vivid sensation of red; I put on the canvas the particular red that satisfies me. A relation is now established between this red and the paleness of the canvas. When I put on besides a green, and also a yellow to represent the floor, between this green and the yellow and the color of the canvas there will be further relations. But these different tones diminish one another. It is necessary that the different tones I use be balanced in such a way that they do not destroy one another. To secure that, I have to put my ideas in order; the relationships between tones must be instituted in such a way that they are built up instead of being knocked down. A new combination of colors will succeed to the first one and will give the wholeness of my conception.
> (Taken from "Art as Experience" by John Dewy, New York, New York: Perigree Books, 1934, p. 126.)

The intensity with which Matisse considers the relationships between colors is similar to the poet's sensitivity to the changing relationships between words and lines in verse, or to the jazz musician's way of balancing various rhythms and tones in an improvisation. If every word a poet utters is part of a living vocabulary, that means it has been said before. A sort of mockingbird, the poet must pluck each word out of the pool of already heavily used vocabulary and "make it new," as Ezra Pound said, by creating patterns of usage that refresh the word by placing it into a context that reframes it—just as Matisse juxtaposed colors making surprising new connections that gave the objects he painted new life.

Pattern making in poetry, or the regularities that persist in any given poet's voice, are not necessarily tied only to metrics, just as form in painting is not always tied to representation. Walt Whitman, our first great American free verse poet, wrote the following poem about the Civil War. What regularities can one spot in the writing that helps us recognize it as poetry?

THE ARMY CORPS

With its cloud of skirmishers in advance,
With now the sound of a single shot snapping like a whip, and now
an irregular volley,
The swarming ranks press on and on, the dense brigades press on,
Glittering dimly, toiling under the sun—the dust-covered men,
In columns rise and fall to the undulations of the ground,
With artillery interspers'd—the wheels rumble, the horses sweat,
As the army corps advances.
(Taken from "The Complete Poems", ed. by Francis Murphy; London, England: Penguin Books, 1975.)

Whitman accumulates images, one by one, causing tension between certain particular things that are happening—the sound of a shot, horses sweating, and wheels rumbling—and the overall swarming of the ranks in general, rising and falling. Each image builds the intensity of the whole scene until we can completely visualize the army corps advancing. The form of the poem is in its consistent language which is "ruled" in the sense that its terminology is imagery. The language of analysis is left out.

Each line of the poem contains one or two distinct and separate images, but those images are also joining together to create an overall coherent scene. The line breaks define the way the reader receives information, bit by bit, controlling the speed with which the reader follows the words down the page. The language restricted to making imagery, one or two completed images per line, and the smaller images piling up to create a much larger scene are all examples of "rules" that a contemporary poem might follow.

The boundary conditions of a poem, or the boundaries establishing the difference between what can be included and what cannot are usually apparent close to the start. One of my favorite poets, Russell Edson, for example, is likely to begin a poem with a sentence that firmly fixes the boundaries.

GRASS

The living room is overgrown with grass. It has come up
around the furniture. It stretches through the dining room, past
the swinging door, into the kitchen. It extends for miles
and miles into the walls.

There's treasure in grass, things dropped or put there, a stick
of rust that was once a penknife, a grave marker...All hidden
in the grass at the scalp of the meadow...

In a cellar under the grass an old man sits in a rocking chair,
rocking to and fro. In his arms he holds an infant, the infant
body of himself. And he rocks to and fro under the grass
in the dark.

(Taken from "The Tunnel: Selected Poems of Russell Edson"; Oberlin Ohio: Oberlin College Press, 2001.)

Very quickly we can spot a few laws that won't be violated without good reason. The voice, for example, won't change its level of diction. It won't turn a corner and sound like a Wallace Stevens poem. Nor will the vision suddenly shift from the surreal to the flatly real. We won't find ourselves suddenly reading an instruction manual. The poem will stick to its narrative style. And it will stick to the humorous tone. It will follow through and won't suddenly leap from its prose-poem shape into tercets. And, finally, the voice will speak in terms of images, not in terms of abstract concepts.

It's true that regularities are sometimes tough to spot in contemporary poems. Rules can disguise themselves by operating in tandem. Or by operating in secret. But all good poetry has "rules" or fundamental strategies that create consistency and unity. The consistency can be accomplished by choosing repetitive methods of creation—and those methods might not be detectable. When I write, for example, I often choose random phrases from such varied sources as travel brochures, driver's education manuals, prayer books, and so forth. Then I try to weave the strangely disparate pieces together, creating a whole. I repeat this process again and again, throwing things out and adding other things, until a poem begins to emerge and an accumulation of images starts to make sense.

Rumor has it that John Ashbery wrote a work with the underlying principle that it had to include the last word taken from the last page of every green book on his shelf. A rule isn't necessarily chosen to be recognized, but it shapes and influences structure. The rule places a constraint on the imagination and, therefore, poses a challenge to the artist.

Without search images for the kinds of consistencies operating in poetry, however, it's no wonder that many readers have trouble distinguishing between the regular and the random in poetry. It's difficult for the average reader to encounter a poem and determine whether it's good or not. A seasoned reader, though, can sense that, in good poetry, subtle laws are being obeyed and that the laws are capable of shaping even frozen accidents into elegant, complex, meaningful and unified structures.

Poets imitate our commonly spoken words and throw them back to us in new patterns to make us hear ourselves in a fresh way. The poet is like the mockingbird! Poets, of course, differ from mockingbirds in the sense that they don't give us exact replicas of what they've heard. They transform the words they hear into constructions that have emotional significance.

Recognizing that there's a strong component of the "found" in poetry, the poet's consciousness acts as a magnet attracting seemingly arbitrary things into its flow. The "arbitrary" things—whether they consist of snatches of overheard conversation, perceptions, or words from a crossword puzzle, can seem as if they're waiting in the wings to be pulled into the stream of the poet's consciousness so that, during its best moments, the poem seems to be effortly emerging about of it's own foundations.

And the mockingbird? A song leaps from its throat, hurtling through space until it falls into someone's circumference of hearing. What does that listener hear? A lot would depend upon his or her level of practice and skill as a listener. One ornithologist, the recently deceased Ted Parker, could identify over 3,000 species of birds by their songs alone. Even the most sensitive listeners of poetry couldn't have skills comparable to Ted Parker's.

Nevertheless, most birders can sense that a combination of both regular and random features are creating the complexity of bird song. A bird song has specific "signatures." that can be detected in a stream of seemingly random notes. Those signatures, which communicate information about territorial boundaries, mating, and feeding have evolved from generation to generation to become increasingly successful.

In the final analysis, if a poem is to be a thing of great depth, beauty, and complexity, it will survive, like bird song, only because it has undergone a rigorous process of evolution. The evolution can take place in the writer's mind or directly on the page. The drafting and revision of any particular poem can take place over a period of days, weeks, or years.

A poem's greatness can also be caused by the context in which it first appears in history—the vernacular, form, or subject of the new creation will seem startling, foreign and new. At the same time the poem will feel deeply connected to a tradition that led up to it.

In general, evolution produces an end result that works, partly because the results that don't work die out. The results produced by evolution, of course, are often very different from one another in terms of form and function. On the surface poetic voice and bird song don't seem at all related, just as the Orangutan of Borneo doesn't appear to have anything in common with the twenty-armed starfish that lives in the sea of Cortez. However, if the rules governing evolution have shaped those differing forms, a profound connection exists between them.

Art and Complexity
J. Casti and A. Karlqvist (editors)

Fractal Expressionism—Where Art Meets Science

Richard Taylor

1 INTRODUCTION

If the Jackson Pollock story (1912–1956) hadn't happened, Hollywood would have invented it any way! In a drunken, suicidal state on a stormy night in March 1952, the notorious Abstract Expressionist painter laid down the foundations of his masterpiece *Blue Poles: Number 11, 1952* by rolling a large canvas across the floor of his windswept barn and dripping household paint from an old can with a wooden stick. The event represented the climax of a remarkable decade for Pollock, during which he generated a vast body of distinct art work commonly referred to as the "drip and splash" technique. In contrast to the broken lines painted by conventional brush contact with the canvas surface, Pollock poured a constant stream of paint onto his horizontal canvases to produce uniquely continuous trajectories. These deceptively simple acts fuelled unprecedented controversy and polarized public opinion around the world. Was this primitive painting style driven by raw genius or was he simply a drunk who mocked artistic traditions? Twenty years later, the Australian government rekindled the controversy by purchasing the painting for a spectacular two million (U.S.) dollars. In the history of Western art, only works by Rembrandt, Velázquez, and da Vinci had commanded more "respect" in the art market. Today, Pollock's brash and energetic works continue to grab attention, as witnessed by the success of the recent retrospectives during 1998–1999 (at New York's Museum of Modern Art and London's Tate Gallery) where prices of forty million dollars were discussed for *Blue Poles: Number 11, 1952*.

Aside from the commercialism and mythology, what meaning do Pollock's swirling patterns of paint really have? Art theorists now recognize his patterns as a revolutionary approach to aesthetics [11, 31]. In a century characterized by radical advances in art, his work is seen as a crucial development. However, despite the millions of words written on Pollock, the precise quality which defines his unique patterns has never been identified. More generally, although abstract art is hailed as a modern way of portraying life, the general public remains unclear about how a painting such as *Blue Poles: Number 11, 1952* (color plate 9) shows anything obvious about the world they live in. The one thing which is agreed upon is that Pollock's motivations and achievements were vastly different from those associated with traditional artistic composition. Described as the "all-over" style, his drip paintings eliminated anything that previously might have been recognized as composition; the idea of having a top and bottom, of having a left and right, and of having a center of focus. Pollock's defense was that he had adopted a "direct" approach to the expression of the world around him, concluding that "the modern painter cannot express this age, the airplane, the atom bomb, the radio, in the old form of the Renaissance...each age finds its own technique" [14].

Appreciation of art is, of course, a highly subjective and personal judgement. Standing in front of one of Pollock's vast canvases, looking at the dense web of interweaving swirls of paint, no one can be told whether such imagery should be liked or not—least of all by a computer. A computer calculates the parameters of an object in a fundamentally different fashion to the human observer. People observe the many different parameters of the painting (for example, the size, shape, texture and color) at the same time, capturing the "full impact" of the painting. In contrast, the computer's approach is reductionist—it separates information, calculating each parameter in isolation. As a consequence, although a computer analysis can never tell you whether an art work should be "liked," it can tell you what the painted patterns "are" with remarkable precision and objectivity. Its reductionist ability to scrutinize individual parameters allows it to quantify information which might have been lost in the "full impact" witnessed by a human observer. The computer can employ its superior computing power (it calculates over six million patterns lying within the canvas) and precision (it examines patterns down to sizes of less than one millimeter) to quantify the painted patterns on Pollock's canvas. This deconstruction of Pollock's paintings into mathematical parameters might, at first, appear to be of little use in the world of art, where human assessments such as beauty, expression and emotion seem more appropriate. However, in this Chapter I will demonstrate that a computer analysis is crucial in order to identify what art theoreticians call the "hand" of Pollock—the trademarks that distinguish him from other artists.

What then is the identifying "hand" of Pollock? The surface of *Blue Poles: Number 11, 1952* has been likened to a battlefield. The vast canvas, stretching across five meters from end to end, contains shards of broken glass embedded in the paint encrusted surface, blood stains soaked into the canvas fabric and eight splattered "poles" violently imprinted by a plank of wood. With each clue shrouded in Pollock mythology, it is clear that an understanding of the essence of Pollock's work requires a rigorous distillation of fact from fiction. During his peak years of 1947–1952, the drip paintings frequently were described as "organic," suggesting the imagery in his paintings alluded to Nature. Lacking the cleanliness of artificial order, his dripped paint clearly stands in sharp contrast to the straight lines, the triangles, the squares and the wide range of other artificial shapes known within mathematics as Euclidean geometry. But if Pollock's swirls of paint are indeed a celebration of Nature's organic shapes, what shapes would these be? What geometry do organic shapes belong to? Do objects of Nature, such as trees and clouds, even have an underlying pattern, or are they "patternless"—a disordered mess of randomness? During Pollock's era, Nature's scenery was assumed to be disordered and his paintings were likewise thought to be random splatters devoid of any order. However, since Pollock's time, two vast areas of study have evolved to accommodate a greater understanding of Nature's rules. During the 1960s, scientists began to examine the dynamics of Nature's processes—how natural systems, such as the weather, evolve with time. They found that these systems weren't haphazard. Although natural systems masqueraded as being disordered, lurking underneath was a remarkably subtle form of order. This behavior was labelled as chaotic and an area of study called chaos theory was born to understand Nature's dynamics [7, 15]. Whereas chaos describes the motions of a natural system, during the 1970s a new form of geometry, called the fractal, was proposed to describe the patterns that these chaotic processes left behind [4, 12]. Since the 1970s many of Nature's patterns have been shown to be fractal, earning fractals the dramatic title of "the fingerprint of God." Examples include coastlines, clouds, flames, lightning, trees and mountain profiles. Fractals are referred to as a new geometry because the patterns look nothing like the traditional Euclidean shapes which humanity has clung to with such familiarity and affection. In contrast to the smoothness of artificial lines, fractals consist of patterns which recur on finer and finer magnifications, building up shapes of immense complexity.

Given that Pollock's paintings often are described as "organic," an obvious step towards identifying the "hand" of Pollock is to adopt the pattern analysis techniques used to identify fractals in Nature's scenery and apply the same process to Pollock's canvases. Following ten years of researching Jackson Pollock, in 1999 I published the results of a computer analysis which revealed that his drip paintings used the same building blocks as Nature's scenery—the fractal [26]. Previous theories attempting to address the artistic significance of Pollock's patterns can be categorized loosely into two related schools of thought—those which consider "form" (i.e., the pattern's significance as a new approach to visual composition) and "content" (i.e., the subject or message the patterns convey). Clearly, the identification of Pollock's patterns as fractal is a vital step for understanding their artistic significance, both in terms of "form" and "content." Rather than using the traditional terminology of Abstract Expressionism, his works are now being re-interpreted as a direct expression of Nature, and the discovery has since been labeled as "Fractal Expressionism" [25, 27].

In this chapter I will discuss the analysis techniques used to identify the fractal fingerprint of Pollock's work. In addition, I will present two recent developments of the research—one which focuses on the "content" and the other on the "form" of his drip paintings. The first concerns the multidisciplinary debate triggered by my results over the precise process that Pollock used to generate his fractal patterns. Exploration of Pollock's painting process raises intriguing questions for many researchers. For art theorists, identification of the method Pollock used to paint the fractals may provide clues as to why he painted them and thus to the artistic meaning—the "content"—of his fractals. His process also offers an intriguing comparison for scientists studying fractal generation in Nature's systems. For psychologists, the process represents an investigation of the fundamental capabilities and limits of human behavior. How did a human create such intricate patterns with such precision, twenty-five years ahead of their scientific discovery? Most examples of "fractal art" are not painted by the artist but instead are generated indirectly using computer graphics [17, 18].[1] How did Pollock construct and refine his fractal patterns? Pollock received significant media attention at his creative peak in 1950 and the resulting visual documentation of his painting technique offers a unique opportunity to study how fractals can be created directly by a human. I will present an analysis of film sequences which recorded the evolution of his patterns during the painting process and I will discuss the results within the context of recent visual perception studies of fractal patterns. Whereas these results explore the generation process in the hope of learning more about the "content" of Pollock's fractals, the second recent development concentrates on the precision with which the "form" of Pollock's fractals can be identified. In response to my results, a number of art museums and private art collectors inquired about the potential of the fractal analysis to authenticate and date Pollock's paintings. As the commercial worth of Pollock's paintings continue to soar, judgements of authenticity have become increasingly crucial. If a new drip painting is found, how do we decide if it is a long-lost masterpiece or a fake? When dealing with such staggering commercial considerations, subjective judgements attempting to identify the "hand" of the artist may no longer be adequate. I therefore will demonstrate the considerable potential that the fractal analysis technique has for detecting the "hand" of Pollock by examining a drip painting which was sent to me to establish its authenticity.

2 THE "DRIP AND SPLASH" TECHNIQUE—A COMPARISON WITH NATURE

Pollock's first exploration of the drip and splash technique took place during the winter of 1942–1943. Described as his "preliminary" phase, he completed the initial stages of the painting using the brushwork style of his previous paintings, but then dripped a final layer of paint over the surface. In late 1945 he moved from Manhattan to the Long Island countryside where he renovated an old barn for his studio and by the end of 1946 his first major drip paintings were under way. The procedure appeared basic. Purchasing yachting canvas from his local hardware store, he often abandoned the European ritual of stretching the canvas. The large canvases simply were rolled out on the floor of the barn, sometimes tacked, sometimes just held down by their own weight. Then he would size the canvas with one or two coats of industrial quality Rivit glue. Even the traditional painting tool—the brush—was not used in its expected capacity: abandoning physical contact

[1]Note that early Chinese landscape paintings also have recently been analyzed for fractal content. See R.F. Voss [32]. Although the individual brushstrokes were found to be fractal, the images constructed from the brushstrokes were non-fractal illustrations. In contrast, for Pollock's paintings the image itself was a fractal pattern.

with the canvas, he dipped a stubby, paint-encrusted brush in and out of a can and dripped the fluid paint from the brush onto the canvas below. The brushes were so stiff that, during a visit to see Pollock at work, fellow artist Hans Hofmann exclaimed, "With this you could kill a man."[2] Sometimes he even wouldn't use a brush, preferring trowels, sticks, or basting syringes. "I continue to get further away from the usual painter's tools," stated Pollock [20]. He used these tools in a strange yet rich variety of ways. A film by Hans Namuth and Paul Falkenberg of Pollock painting shows him sometimes crouched down near the canvas, almost drawing with the drips. Other actions show him flinging the paint across large distances. William Rubin (a previous director of The Museum of Modern Art, New York) described another variation on the dripping process, where Pollock would "place a stick in the can of paint, and by tilting the can, let the pigment run down the stick onto the canvas" [23]. Sometimes he even poured the paint directly from an open can. All of these techniques—which abandoned contact with the canvas—are grouped loosely under the descriptive label "drip and splash" technique.

In terms of the paint, an industrial quality enamel called Duco (technically, pyroxylin lacquer), was found to be the most pliable. Pollock had first used Duco in the 1930s during art workshops, and he now returned to using it because of its ease of application, good covering power and quick drying qualities. In addition, aluminum, silver, gouache and oil paints all contributed to his growing battery of effects. The stream of paint produced a continuous trajectory and its character was determined by physical and material variables refined and mastered by Pollock, including the viscosity of the paint, and the height, the angle and the speed of pouring. Most crucially, the paint trajectory directly mapped the artist's gestures and movements. Pollock's wife, Lee Krasner, later remarked that the secret to Pollock's success was his ability to work in air and know exactly where the paint would land—the paint trajectories captured two-dimensional fingerprints of his three-dimensional motion through the air. One of Pollock's friends, Bob Friedman, wrote, "Once Pollock painted in the barn his work began to open up, to shout, to sing" [5]. The barn offered freedom of motion, allowing him to approach his vast canvases from all four sides. Only afterward did he decide which way was up and which was down. As Pollock explained, "On the floor I am more at ease. I feel nearer, more a part of the painting, since this way I can walk around it, work from the four sides and literally be *in* the painting" [20]. Figure 1 shows a photograph of Pollock above one of his paintings.

Contrary to mythology, a Pollock painting was not born easily. Pollock revisited his canvases frequently over weeks or even months, building up intricate layers woven into the increasingly dense web of paint. A typical canvas would be worked and reworked many times. Although he preferred not to think of his approach in terms of stages, the tortuous history of many of his paintings followed a common route. On his own admission, he would start the process by picking the paint can which was nearest at hand. Working feverishly, he would apply paint quickly in short, decisive bursts and the canvas would be covered with a basic pattern within half an hour. Namuth recalls photographing this first stage: "His movements, slow at first, gradually became faster and more dance-like as he flung black, white and

[2]Hoffmann's remark is quoted in O'Connor [14, p.203].

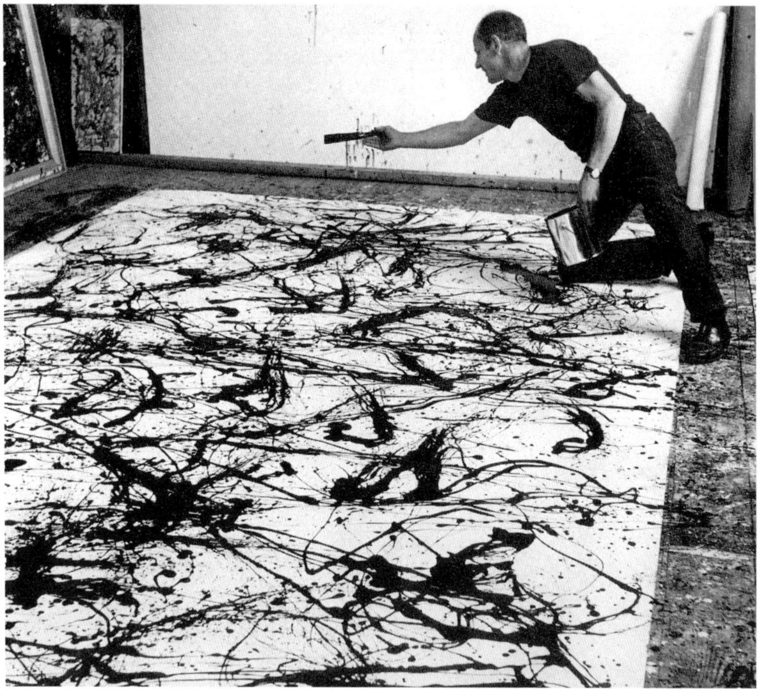

FIGURE 1 A photograph of Pollock with *Number 32, 1950* taken in 1950 by Rudolph Burckhardt. (© 2002 The Estate of Rudolph Burckhardt.) *Number 32, 1950* (enamel on canvas, 269 by 457.5cm) was painted by Pollock earlier in 1950 (Kunstsammlung Nordrhein-Westfalen, Dusseldorf, Germany). (© 2002 The Pollock-Krasner Foundation/Artists Rights Society (ARS), New York.)

rust-colored paint onto the canvas. My photography session lasted as long as he kept painting, perhaps half an hour. In all that time, Pollock did not stop" [13]. After this initial session, Pollock then would break off and step back for a period of contemplation and study. The first stage was over and at this point Pollock would have little idea when he would be sufficiently inspired to resume the painting. Instead, his thoughts would turn to other paintings. In this way, Pollock kept three or four paintings open at all times. However, no painting was safe—a canvas which had laid dormant for several months suddenly would regain his attention. Subsequently, a getting acquainted period would ensue, in which he decided how to add strength to the pattern. The next burst of activity might range from a small adjustment to a complete re-working. Glancing up from a newspaper he might spy some "imperfection" in the pattern and reach awkwardly around the back of his chair, "correct it" with whatever brush was closest at hand and then return to the paper. Alternatively, the painting might be deemed to be lost and he would start from scratch. In this way, the periods between different assaults could vary vastly from work to work. Some of Pollock's layers were laid down quickly, "wet on wet." In other works, a layer would be scrubbed into the surface with a rag before the next layer was deposited. Sometimes, Pollock would wait for a layer to dry and the following day the painting would be nailed to the studio wall for further contemplation (although the extent to which this happened remains controversial). Some paintings were even stretched before being reworked. To the casual observer the whole approach appeared very random. But, in this unique way, Pollock would revisit a painting at irregular periods until he decided the pattern was complete—"concrete" in his language—and the painting would be a record of his released experience.

There are a number of striking similarities between the above description of Pollock's painting style and the processes used by Nature to build it's scenery. His cumulative painting process is remarkably similar to the way patterns in Nature arise—for example, the way leaves fall day after day to build a pattern on the ground, or the way waves crash repeatedly on the shore to create the erosion patterns in the cliff face. The variation in intensity of his painting process also mirrors Nature: just as, say, the rain might change from a short, light drizzle on one day to an extensive storm on the next, Pollock's sessions would vary from small corrections on one day to major re-workings on the next. His method of leaving a painting dormant for a while before his return for a new onslaught is also similar to the cyclic routine of Nature—for example, the tides and the seasons. More generally, this idea of the painting as an ongoing process, occurring indefinitely, clearly reflects Nature's process of pattern generation. Out of all the Abstract Expressionists, Pollock was one of the main pursuers of this so-called "continuous dynamic" process. Another was Arshile Gorky, whose explanation is very similar to Pollock's: "When something is finished that means it is dead, doesn't it? I believe in everlasting. I never finish a painting—I just stop working on it for a while. I like painting because it is something I never come to the end of."[3] "He hated signing. There's something so final about the signature," recalled Krasner about Pollock.[4] Clearly, Nature has an advantage over the artist—its process really can go on forever while Pollock had practically-induced time frames. This notwithstanding, the "continuous dynamic" painting process suggests a closeness with natural evolution. Pollock's unease with the signing process—the artificial act which recognizes the canvas as an "art work" rather than as a piece of Nature—suggests an empathy that Pollock may have felt for natural processes. Furthermore, Pollock's spontaneous and unpremeditated painting process also bears a striking similarity to Nature: Nature doesn't prepare and think about its patterns—they are determined by the interaction with the environment at the specific moment in time that the patterns are being created. Interestingly, because of his unpremeditated style, Pollock often was regarded as an Action Artist and during Pollock's era the art critic Harold Rosenberg said that Action Art had "broken down every distinction between art and life," and that "the painting is not Art; it's an Is. It's not a picture of a thing: it's the thing itself...it doesn't reproduce Nature; it is Nature" [22]. Other parallels between Pollock's method of painting and natural processes are apparent. Gravity plays a central role for both Pollock and Nature.

[3]Gorky's statement was originally quoted by Talcott Clap, *Arshile Gorky: Paintings, Drawings, Studies* 43, catalogue. The Museum of Modern Art, New York, 1962.

[4]Krasner is quoted in Friedman [5, p.185].

Pollock's whole approach of abandoning the easel and instead laying the canvas on the ground triggers comparisons with Nature's processes. Many of Nature's patterns are built in the horizontal plane and controlled to an extent by gravity (for example, the falling rain or falling leaves). Unlike traditional brushstroke techniques, Pollock's dripping technique similarly exploited gravity. Furthermore, in adopting the horizontal plane, the canvas became a physical space, a terrain to be traversed.

Just as there are similarities between the processes used by Nature and Pollock, there are also similarities between the patterns generated by these processes. Pollock abandoned the heavy frames that in previous art movements had been used to isolate the work from its surroundings. His whole philosophy of what he called an "unframed space" was, indeed, compatible with Nature—no natural scenery has its patterns artificially bound and restricted. Some of Pollock's paint trajectories ignore the artificial boundaries of the canvas edge and travel beyond it—a characteristic which is alien to traditional artistic compositional values and clearly closer to Nature's expansive and unconfined patterns. Pollock's choice of a large canvas—one which dominates a viewer's environment—is also similar to Nature's scenes. As fellow Abstract Expressionist Mark Rothko described in 1951, "To paint a small picture is to place yourself outside the experience...However, you paint the larger picture, you are in it."[5] Similarly, Pollock regarded his own vast paintings as environments. Furthermore, his approach from all four sides of the canvas replicated the isotropy and homogeneity of many natural patterns. The resulting uniformity of his "all-over" composition lacks any center of focus. This characteristic stands in contrast to traditional art compositions and instead shows similarities with Nature's patterns. Like natural patterns, Pollock's paintings are also astonishing feats of pictorial invention: Nature's processes and the painting process chosen by Pollock both generate a rich variety of complex structure. No two natural patterns are exactly the same and this is also true of Pollock's paintings. Color plates 10 and 11 compare some typical natural patterns with those of Pollock's drip trajectories. Although there are specific differences, an underlying, shared quality can be identified. Could it therefore be that the basic trademark of Nature's pattern construction also appears in Pollock's drip work? The superficial similarities between both the processes and patterns used by Pollock and Nature clearly offer clues to a common approach. However, in order to establish this connection rigorously, it is first necessary to go beyond vague descriptions such as "organic" and, instead, identify the precise generic qualities of Nature's dynamics and the patterns produced.

3 CHAOS

In 1960, Edward Lorenz used a computer to plot the irregularities in weather patterns. His findings, which could be applied to many of Nature's processes, showed the weather to be fundamentally unpredictable. He found a new regime

[5] Mark Rothko's statement originally appeared in *Interiors*, May 1951 and is quoted by Charles Harrison, 'Abstract Expressionism', *Concepts Of Modern Art*, edited by Nikos Stangos, 196. London: Thames and Hudson, London, 1974.

of behavior—one which would be later labeled as chaos [7, 15]. When he looked carefully at how the weather varied with time he found that, despite apparent randomness, the behavior was really an intermediate state between highly ordered behavior and fully random behavior. He noticed that the behavior had a pattern, but with disturbances—a kind of orderly disorder. Unlike truly ordered systems, the disturbances made the behavior unpredictable. But it wasn't fully random behavior either. There was an underlying quality—an order which was masquerading as randomness. He found that the defining characteristic of chaotic systems lay in the equations which mapped out their behavior. The equations are not like those of well-ordered systems, where the outcomes are relatively insensitive to small changes in the parameters fed into the equations. Well-ordered systems are predictable because a small error in the knowledge of the initial conditions will not alter greatly the way the system evolves with time. In contrast, the outcomes of chaotic equations are exponentially sensitive to initial conditions. Tiny differences in the starting conditions become magnified as the system evolves, resulting in exponentially diverging outcomes and hence a long term behavior which cannot be predicted. This signature of chaos (extreme sensitivity to initial conditions) often is referred to as the Butterfly Effect—where even a small variation, such as a butterfly flapping its wings in the Amazon, could have dramatic consequences for the wind patterns across the North American skies. The sensitivity of Nature's chaos arises in part due to a principle called "holism"—where everything in the system is connected intimately to, and hence sensitive to, everything else.

Since its discovery, chaos theory has experienced spectacular success in explaining many of Nature's processes. Given the similarities with Nature, could Pollock's painting process therefore also have been chaotic? There are two revolutionary aspects to Pollock's application of paint and, remarkably, both have potential to introduce chaos. The first is his motion around the canvas. In contrast to traditional brush-canvas contact techniques, where the artist's motions are limited to hand and arm movements, Pollock used his whole body to introduce a wide range of length scales into his painting motion. In doing so, Pollock's dashes around the canvas possibly followed Lévy flights: a special distribution of movements, first investigated by the mathematician Paul Lévy in 1936, which has recently been used to describe the statistics of many natural chaotic systems.[6] The second revolutionary aspect concerns his application of paint by letting it drip on to the canvas. In 1984, a scientific study of dripping fluid by Robert Shaw at the University of California showed that small adjustments to the "launch conditions" (in particular, the initial flow rate of the fluid) could change the falling fluid from a non-chaotic to chaotic flow [24]. Although Shaw considered the case of a flow which occurs in drips, the result is true also for a continuous flow of liquid. For example, water flowing along a river or a pipe can be made to descend into the turbulence of chaos by simply adjusting the way the liquid is "launched."

[6] Whereas random motion is described by Brownian statistics, chaotic motion can be described by Lévy statistics. In Brownian motion a particle makes random jumps (or 'flights') and each jump is usually small: the resulting diffusion can be described by a Gaussian distribution with a finite variance. In Lévy diffusion, on the other hand, long jumps are interspersed with shorter jumps, and the variance of the distribution diverges. For more details on Lévy flights see Klafter et al. [9] and C. Tsallis [30].

Whether working with a dripping or continuos flow, it is possible to tune the flow from chaotic to non-chaotic and vice versa. Therefore, Pollock likewise could have mastered a chaotic flow of dripping paint. Whether he was dripping fluid from a stick or directly from a can, he could have mastered a particular "launch" condition, for example a particular flick of the wrist, which generated chaos in the falling fluid.

Perhaps, then, this potential for chaotic activity could be confirmed by an analysis of the dynamics of these two sets of Pollock's motions? In 1950, Hans Namuth took a series of black and white photographs and together with Paul Falkenberg shot a 16mm color film of Pollock painting. However, although suggestive, the film footage lacks the statistics required to confirm conclusively that the motions in his two processes were exclusively chaotic. To obtain reliable statistics, it would be necessary to undertake a detailed examination (including simultaneous zoom shots from above and the side) of the motions of Pollock and his dripping paint during production of at least one hundred trajectories. Even if Pollock were alive today to participate in such an exhaustive experiment, the filming conditions would be sufficiently intrusive to make the approach impractical. Therefore, an analysis of his painting motions is destined to be a limited method for deducing if he was chaotic. Instead, it is more informative to look for the signature of his chaotic activity in the patterns which record the process—the paint trajectories themselves. The logical way forward in this investigation is to make a visual comparison of the drip trajectories painted by Pollock and drip trajectories generated by processes which are known to be chaotic. If Pollock's trajectories are likewise generated by chaotic motion, then the resulting trajectory patterns should have similar visual characteristics to those of the known chaotic system.

The two chaotic processes proposed for generating Pollock's paint trajectories occur over distinctly different size ranges. These sizes can be estimated from the film and still photography of Pollock's painting process. Based on the physical range of his body motions and the canvas size (which, for example, is 4.8m for *Blue Poles: Number 11, 1952*), the Lévy flights in his motion across the canvas are expected to have had lengths with values lying approximately between 1cm and 4.8m. Thus, his Lévy flights would have generated features in the resulting paint trajectories with sizes lying between the same values. In contrast, the drip process is expected to have shaped the paint trajectories over much finer sizes—between the approximate sizes of 1mm and 5cm. This range is calculated from variables which affect the drip process (such as paint velocity and drop height) and those which affect paint absorption into the canvas surface (such as paint fluidity and canvas porosity). As a result, Pollock's motion around the canvas predominantly influenced the directions of the paint trajectories—what I will call the "bones" of the trajectories. In contrast, the dripping process predominantly determined what I will call the "flesh"—the variations in the thickness of the trajectories. Figure 2 shows a schematic representation of this concept. The top pattern shows the trajectory "bones" shaped by Pollock's motion around the canvas. The bottom pattern shows both the "skin" and the "bones," generated by contributions from both Pollock's motion and the drip motion.

First I will investigate the trajectory bones produced by the motion of Pollock across the canvas. A simple system which generates drip trajectories can be de-

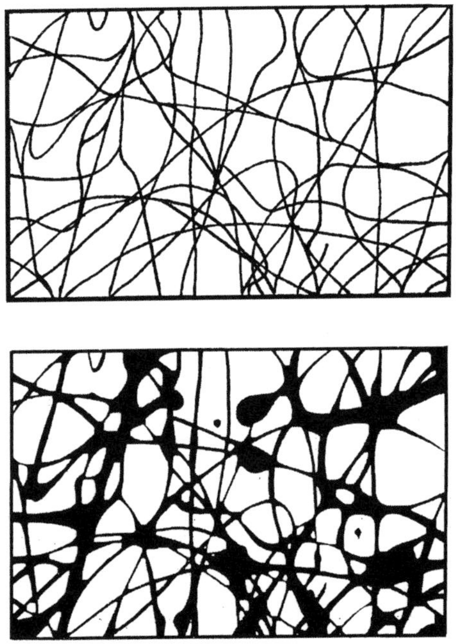

FIGURE 2 A schematic demonstration of the "flesh" and "bones" of Pollock's trajectories. The bottom picture shows the "skin" and "bones" of the trajectories. The top picture shows only the "bones." The "flesh" is produced by the dripping process while the "bones" are produced by Pollock's motion around the canvas. These trajectories are taken from a 13cm by 20cm section of *Number 14, 1948*. *Number 14, 1948* (enamel on gesso on paper, 57.8cm by 78.8cm) was painted by Pollock in 1948 (Yale University Art Gallery, New Haven). (© 2002 The Pollock-Krasner Foundation/Artists Rights Society (ARS), New York.)

signed where the degree of chaos in this motion can be tuned. A series of patterns then can be produced which vary from non-chaotic through to totally chaotic. In fact, the ideal system to do this was introduced to the art world in its crudest form by Ernst in 1942 as he dripped paint from a bucket. Ernst, who was continually on the lookout for new artistic techniques, described his procedure as follows: "Tie a piece of string, one or two meters long, to an empty tin can, drill a small hole in the bottom and fill the tin with fluid paint. Then lay the canvas flat on the floor and swing the tin backward and forward over it, guiding it with movements of your hands, arms, shoulder and your whole body. In this way surprising lines drip onto the canvas."[7] What Ernst did not realize, and nor did the scientists of that era, was that the unguided pendulum was an even more interesting system than the guided one. When left to swing on its own, the container follows an el-

[7]Max Ernst is quoted by Werner Schmalenbach, *Masterpieces Of 20th Century Art*, 254. Munich, Prestel-Verlag, 1990.

lipse which spirals into the center, and this motion is captured by an identical trajectory pattern on the canvas. This is a well-defined motion (called damped simple harmonic motion) and the pattern generated is non-chaotic. It is due to this very stable, predictable motion that pendula are found in the clocks of many households around the world. This periodicity in the pendulum's motion is one of science's legendary discoveries: Galileo Galilei made his discovery by swinging a church lamp back and forth. And this is how the story stayed for more than three hundred years until the 1980s, when chaologists surprised the scientific community by discovering that the very same pendulum, this great symbol of regularity, also could be made to descend into the unpredictability of chaos [29]. By knocking the pendulum at a frequency slightly lower than the one at which it naturally swings, the system changes from a freely swinging pendulum to what is called a "forced pendulum" or "kicked rotator." The motion of this system now has become a standard system in science for demonstrating chaotic trajectories: by tuning the frequency and size of the kick applied to the pendulum, the motion can evolve from the non-chaotic motion of the free pendulum through to increasingly chaotic motion.

Using a refinement of this pendulum concept, I have generated the two distinct categories of drip trajectory paintings [28]. Figure 3 shows a drip painting in progress. Example sections of non-chaotic (top) and chaotic (middle) drip paintings are shown in figure 4. Since Pollock's paintings were built from many criss-crossing trajectories, these pendulum paintings likewise feature a number of trajectories generated by varying the pendulum's launch conditions. For comparison, the bottom picture is a section of Pollock's *Number 14, 1948* painting. Introduction of chaos into the pendulum's motion induces a clear evolution in the "bones" of the drip trajectories and, in a visual comparison, Pollock's trajectories bear a closer resemblance to the pattern generated by the chaotic motion (middle) than to the non-chaotic motion (top). Figure 4, therefore, offers a clue that Pollock's motion around the canvas was chaotic. In fact, the introduction of chaos into the motion of the pendulum induces a change in both the "flesh" and the "bones." The sharper changes in trajectory direction within the "bone" pattern of the chaotic system also induces variations in the dynamics of the dripping fluid. This generates the large variations in the "flesh" thickness across the painting observed in figure 4 (middle). This result, therefore, emphasizes an inter-dependence of the two processes determining the "flesh" and the "bones": the dynamics of the container's motion can affect the dynamics of the falling fluid. This is expected also for Pollock's painting technique—his motions around the canvas will have contributed to the way he launched the paint from his painting implement. Nevertheless, the motion of the launch implement (whether it was a stick or a can) is just one of the factors which dictates the launch conditions and hence the dynamics of the falling liquid. For example, it is possible to hold the launch implement stationary, so removing the issue of its motion, and adjust other launch parameters to tune the falling fluid from a chaotic to non-chaotic flow. In particular, the falling liquid's flow rate and viscosity can be adjusted. In Shaw's original experimental investigation of the chaotic dynamics of dripping fluid, he used a stationary tap and adjusted its aperture to change the flow rate of the fluid. A laser beam was shone through the falling fluid and used to detect the emergence of chaos as the dynamics of the fluid flow was altered. Figure 5 shows the pattern created by a similar experimental set-up where I used a chaotic flow of paint to create a pattern on a horizontal canvas placed below the tap. In figure 5 this drip pattern generated by the chaotic flow of paint is compared to the "flesh" of Pollock's paint trajectories. Again, there is a visual similarity. This offers a clue that Pollock mastered a method of launching paint from his painting implement which induced chaos in the motion of the falling fluid.

The preliminary investigations shown in figures 4 and 5, therefore, raise the possibility that the two components of Pollock's painting process—his motion around the canvas and his dripping process—both produced similar visual qualities to patterns generated by chaos. If Pollock's drip patterns were generated by chaos, what common quality would be expected in the patterns left behind? Many chaotic systems form fractals in the patterns that record the process [4, 7, 12, 15]. Is the shared visual quality revealed in figures 4 and 5, therefore, the fractal? It should be stressed that the above comparisons serve only as initial indications. Both Pollock's patterns and those generated by chaos are characterized by endless variety, and thus a comparison of two individual patterns is of limited use. Furthermore, any similarities detected by using only two pictures might be a result of coincidence rather than a sign of a common physical origin. Thus, although important as initial clues to fractal content, a systematic analysis involving an assessment of many patterns must be undertaken. Furthermore, the analysis should be divorced from subjective visual comparisons and instead should involve the calculation of a parameter which identifies and quantifies the fractal content objectively. An analysis which does this will be presented in the next section.

4 FRACTALS

In 1975 Benoit Mandelbrot coined the term "fractal" from the Latin adjective to mean "fractured" or "irregular" [4, 12]. Although appearing fractured and irregular during a superficial inspection, a more detailed examination of Nature's fractals reveals a subtle form of repeating order. Mandelbrot showed that Nature's fractal patterns obey a scaling relationship called statistical self-similarity: the patterns observed at different magnifications, although not identical, are described by the same statistics. The results are visually more subtle than the instantly identifiable, artificial fractal patterns generated using exact self-similarity, where the patterns repeat exactly at different magnifications. However, there are visual clues which help to identify statistical self-similarity. The first relates to "fractal scaling." The visual consequence of obeying the same statistics at different magnifications is that it becomes difficult to judge the magnification and hence the length scale of the pattern being viewed. This is demonstrated in color plate 12(top) and plate 12(middle) for Nature's fractal scenery and in plate 12(bottom) for Pollock's painting. A second visual clue relates to "fractal displacement," which refers to the pattern's property of being described by the same statistics at different spatial locations. As a visual consequence, the patterns gain a uniform character and this is confirmed for Pollock's work in figure 6, where the pattern density P is plotted as a function of position across the canvas.

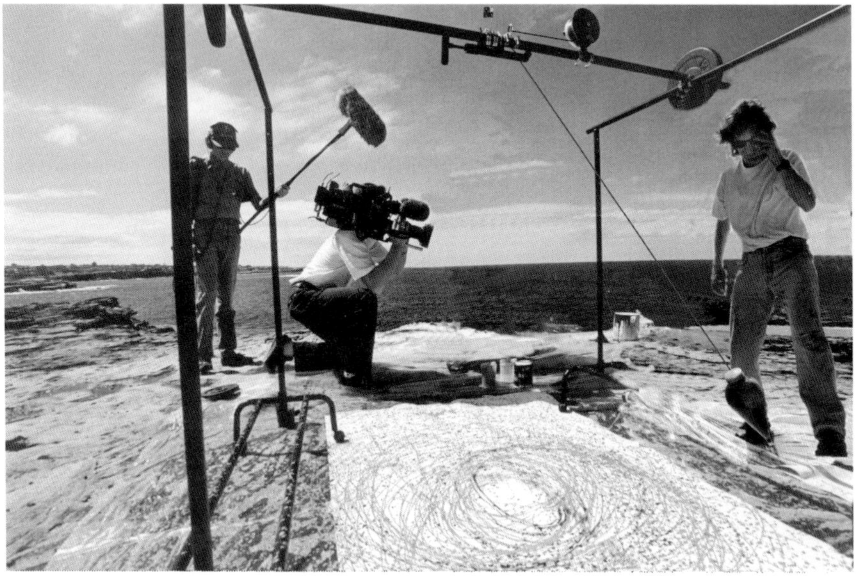

FIGURE 3 A photograph of the guided pendulum apparatus (referred to as the "Pollockiser") used to generate drip paintings. The photograph was taken during filming for the Australian Broadcasting Corporation's television program *Quantum* in March 1998.

Pollock's patterns therefore display both "fractal displacement" and "fractal scaling"—the two visual clues to fractal content. When Mandelbrot first introduced the concept of fractals, he stressed that perhaps the most significant associated advance was that there was now a way of identifying and describing patterns which were previously beyond scientific quantification. He noted, "Scientists will...be surprised and delighted to find that now a few shapes they had to call grainy, hydra-like, in between, pimply, pocky, ramified, seaweady, strange, tangled, tortuous, wiggly, whispy, and the like, can henceforth be approached in rigorous and vigorous fashion" [4, 12]. I will now apply this "rigorous" approach to Pollock's trademark patterns to confirm the visual clues to fractal content. A traditional method for detecting statistical self-similarity is shown in figure 7 for a schematic representation of a Pollock painting. A digitized image (for example a scanned photograph) of the painting is covered with a computer-generated mesh of identical squares. By analyzing which squares are occupied by the painted pattern (shaded gray in figure 7) and which are empty, the statistical qualities of the pattern can be calculated. Reducing the square size is equivalent to looking at the pattern at a finer magnification. Thus, in this way, the pattern's statistical qualities can be compared at different magnifications. A crucial parameter in characterizing a fractal pattern is the fractal dimension, D, and this quantifies the scaling relationship between the patterns observed at different magnifications [4, 8, 12]. For Euclidean shapes, dimension is a simple concept and is described by the familiar integer values. For a smooth line (containing no fractal structure) D has a value of

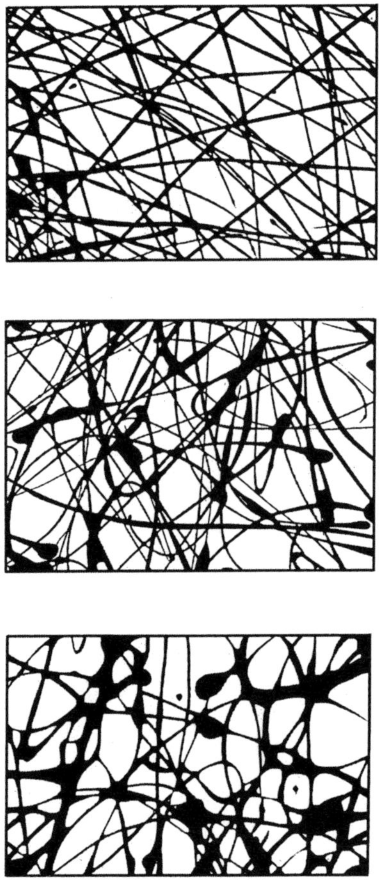

FIGURE 4 The detailed patterns of non-chaotic (top) and chaotic (middle) drip trajectories generated by the "Pollockiser," compared to a section of Pollock's *Number 14, 1948* painting (bottom). All three sections are approximately 13cm by 20cm. (© 2002 The Pollock-Krasner Foundation/Artists Rights Society (ARS), New York.)

1, while for a completely filled area its value is 2. However, for a fractal pattern, the repeating structure at different magnifications causes the line to begin to occupy area. D then lies in the range between one and two and, as the complexity and richness of the repeating structure increases, its value moves closer to 2. D can be used therefore to identify and quantify the fractal character of a pattern. Using the computer-generated mesh shown in figure 7, D can be obtained by comparing the number of occupied squares in the mesh, $N(L)$, as a function of the size, L, of the squares. For fractal behavior, $N(L)$ scales according to the power law relationship $N(L) \sim L^{-D}$, where D has a fractional value lying between 1 and 2 [4, 8, 12]. Therefore, by constructing a "scaling plot" of $\log N(L)$ against $\log L$ the fractal

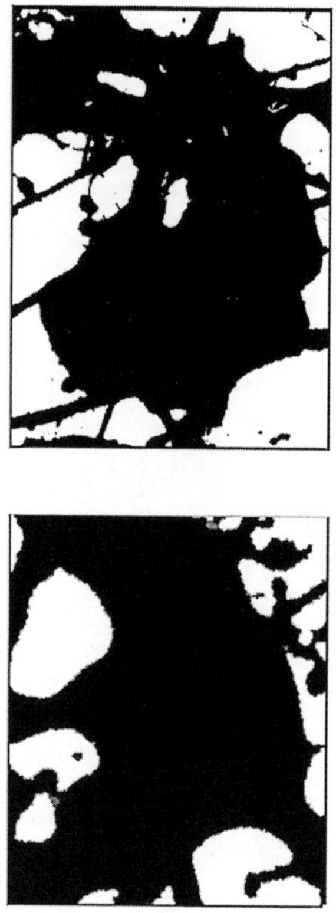

FIGURE 5 A comparison of the paint marks (the "flesh" pattern) made by Pollock's drip process (top) and those produced by a known chaotic flow (bottom). The top image is taken from a section of Pollock's *Number 32, 1950*. Both sections are approximately 13.5cm by 18cm.

behavior manifests itself as the data lying on a straight line and the value of D can be extracted from the gradient of this line.

The two chaotic processes proposed for generating Pollock's paint trajectories operated across distinctly different length scales. His chaotic Lévy flights across the canvas are expected to have occurred over relatively large distances—between 1cm and 4.87m (where 4.87m corresponds to the canvas size). In contrast, the chaotic drip process is expected to have shaped the trajectories over significantly smaller sizes—between 1mm and 5cm. Due to the presence of these two chaotic processes, it is expected that the fractal analysis of Pollock's paintings will reveal the presence of two sets of fractals patterns—one set occurring across small L values and one set across large L values. Furthermore, because the two chaotic processes have distinctly different physical origins (one generated by drips and the other by Pollock's motions) the two sets of fractal patterns are expected to be described by different D values. Note that systems described by two or more

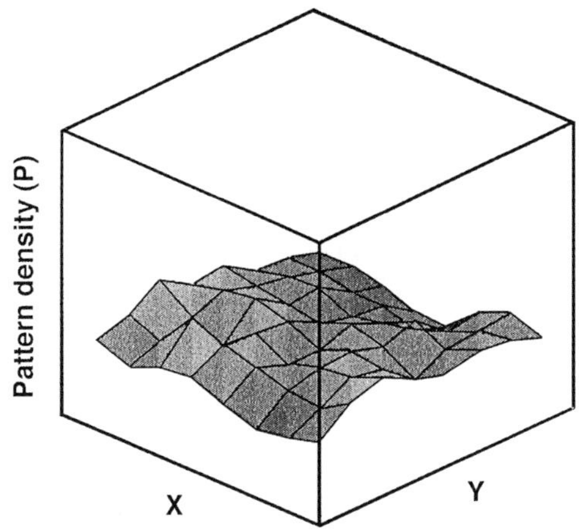

FIGURE 6 A plot showing that the pattern's spatial density P remains approximately constant at different locations across the canvas. To produce this plot, the scanned photograph of a Pollock painting is covered with a computer-generated mesh of identical squares. Within each square, the percentage of the canvas surface area covered by the painted pattern (P) is then calculated. The X and Y labels indicate the locations in the length and height directions respectively. The square size used to calculate P is $L = 0.05$m. The plot is for the painting *Number 14, 1948*. The plotted ranges are as follows: P spans the range between 0 and 100%, X and Y span the range from 0 to 0.43m.

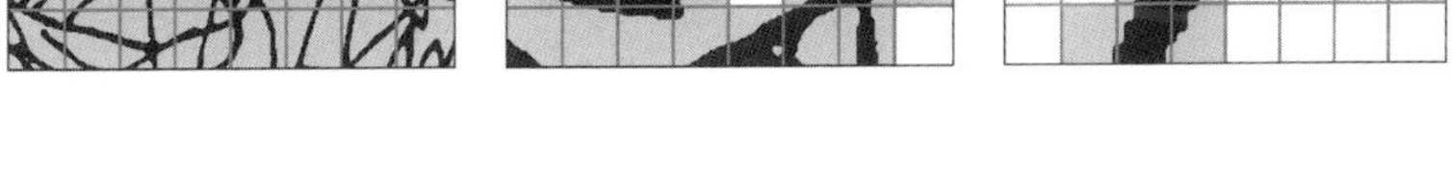

FIGURE 7 A schematic representation of the technique used to detect the fractal quality of Pollock's patterns. A computer generated mesh of identical squares covers the surface of the painting. Then the size of the squares in this mesh is decreased gradually. From left to right the square size is decreased and in each case the number of "occupied" boxes (indicated by the gray shading) is counted.

D values are not unusual: trees and bronchial vessels are common examples in Nature. I will label the value of L at which the transition between the two sets of fractal patterns occurs as L_T. Based on the size ranges specified above, L_T is expected to occur at approximately 1cm.

The analysis of Pollock's paintings confirm these expectations. For example, figure 8 shows the "scaling plot" for the aluminum paint trajectories within *Blue Poles: Number 11, 1952* which has a canvas size of 2.10m high by 4.87m wide. Data plotted on the extreme left hand side of the graph corresponds to patterns with a size $L = 0.8$mm while data on the extreme right hand side corresponds to $L = 10$cm (note, although not shown, the analysis continues beyond the range shown, right up to $L = 1$m). The data, represented by the black line, follows one gradient for small L values (i.e., on the left) and another gradient for large L values (on the right). The different gradients indicate that the two fractal patterns have different D values. These two D values are labeled as the drip fractal dimension, D_D, (produced by the chaotic motion of the dripping paint) and the Lévy flight fractal dimension, D_L, (produced by the chaotic motion of Pollock's Lévy flights across the canvas). For the aluminum paint trajectories analyzed in figure 8, the fractal dimensions have values of $D_D = 1.63$ and $D_L = 1.96$. The value of L_T (the size at which the transition between the D_D and D_L ranges occurs) is 1.8cm, consistent with the value predicted above. The patterns were analyzed for L values ranging from 1mm up to 1m and fractal behavior was observed over this complete range—the largest observed fractal pattern in the painting is over one thousand times larger than the smallest. This immense size range is significantly larger than for observations of fractals in other typical physical systems.[8] One of the consequences of observing the fractal patterns over such a large size range is that the fractal dimension can be determined with great accuracy.

Having established their fractality, how do these fractal patterns evolve in character through the years? Art historians categorize Pollock's development of the drip technique into his "preliminary" phase (circa 1943), his "transitional" phase (circa 1947) and his "classic" phase (circa 1950). The sparse drip patterns of his "preliminary" drip paintings were deposited over a foundation of brushed paint (see, for example, *Water Birds*, shown in color plate 13). To determine the fractal quality of the drip trajectories of these paintings, the underlying layers of brushed paint were removed electronically from the scanned images of the paintings prior to the analysis. For these "preliminary" paintings the dripped layer was found to have a very low fractal dimension and that the foundation patterns of brushed paint were not fractal. For Pollock's "transitional" paintings, the dripped layers of paint assumed a more dominant role over the underlying brushed layers in regard to their contribution to the visual impact of the painting (see, for example, *Full Fathom Five*, shown in color plate 10). For these "transitional" paintings the dripped layers were found to have a higher fractal dimension than for the dripped layer of his "preliminary" paintings. Furthermore, the patterns established by the

[8]Unlike fractal patterns generated by mathematical equations, fractals in physical systems do not range from the infinitely large through to the infinitesimally small. Instead physical fractals are observed across only a limited range of sizes. A recent survey of observations of fractals in physical systems suggests that the largest pattern is typically only 30 times larger then the smallest pattern. See Avnir et al. [3].

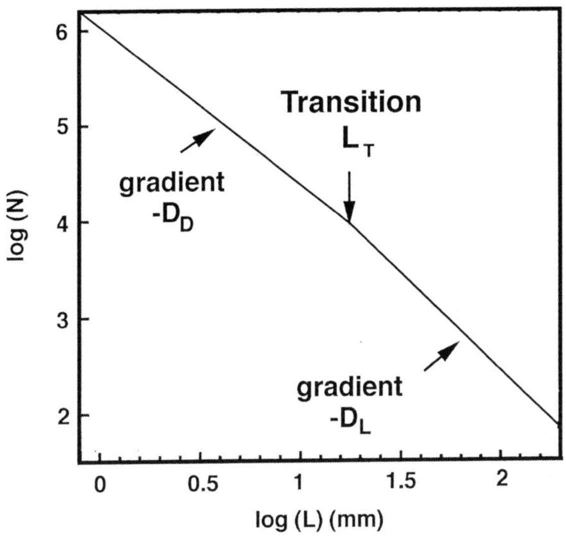

FIGURE 8 A plot of $\log(N)$ versus $\log(L)$ for the pattern created by the aluminum paint trajectories within *Blue Poles: Number 11, 1952.* The horizontal axis spans the range between 0.8mm and 10cm. The data points lie on the black line—for clarity, due to the large number of points (over 2000), the individual points are not shown.

underlying brushed layer (with the dripped layers removed) was also fractal. For Pollock's "classic" drip paintings, there were few, if any, underlying brush marks—the paintings were constructed almost entirely from layers of dripped paint (see, for example, *Blue Poles: Number 11, 1952* shown in color plate 9). For these paintings, the dripped layers were found to have even higher fractal dimensions than the dripped layers of the "transitional" paintings.

Consider the evolution of the D_D and D_L values in more detail. The analysis shows that D_D gradually increased over the years. The drip patterns of *Untitled: Composition With Pouring II* and *Water Birds*, both painted in 1943, have low D_D values. Similarly, *Untitled*, painted in 1945 has a D_D value of 1.12. This indicates that the drip trajectories within these "preliminary" paintings have low fractality for the small sizes of L characterized by D_D. However, by 1946, when Pollock painted *Free Form*, he had succeeded in refining his drip technique to produce well-defined fractal patterns. *Number 14, 1948* painted in 1948, has a D_D value of 1.45 and *Autumn Rhythm: Number 30, 1950* painted in 1950, has a D_D value of 1.67. In 1951 Pollock mostly used the drip technique to draw figurative representations and these paintings are not fractal. However, he still occasionally painted non-figurative "all-over" compositions with the drip technique, such as *Untitled* with a D_D value of 1.57. In 1952, *Blue Poles: Number 11, 1952* represented a final and brief return to his "classic" style and has the highest D_D value of any completed Pollock painting with a value of 1.72.

TABLE 1 The results of the fractal analysis, revealing an increase in fractal dimension through the years 1945–1952.

Year	Painting Title	D_D	D_L	Canvas Area (m^2)	A (%)
1945	*Untitled*	1.12	–	0.24	4
1947	*Lucifer*	1.64	>1.9	2.79	92
1948	*Number 14, 1948*	1.45	>1.9	0.46	28
1949	*Number 8, 1949*	1.51	>1.9	1.56	86
1950	*Number 32, 1950*	1.66	>1.9	12.30	46
1950	*Autumn Rhythm*	1.67	>1.9	14.02	47
1951	*Untitled*	1.57	>1.9	0.53	38
1952	*Blue Poles*	1.72	>1.9	10.22	95

The fractal quality of his patterns at large L sizes—as characterized by D_L—also evolved through the years. Analysis of the patterns of *Untitled: Composition With Pouring II, Water Birds* (1943), and *Untitled* (1945) do not yield straight lines in the region of the graph where D_L should be extracted, indicating that these early paintings are non-fractal in the large L size regime (just as they are not fractal in the small L size regime). However, by the time he painted *Free Form* in 1946 Pollock had evolved his drip paintings such that they had become fractal for the large L sizes. Indeed, throughout the period 1947–1952 (excluding 1951; see above), the D_L values are significantly higher than the D_D values. In other words, as Pollock perfected his technique, the fractal patterns at the large L sizes became significantly more dense with fractal structure than the fractal patterns at small L sizes. In fact, during his "classic" period of 1950 the D_L values became remarkably high, approaching a value of 2. This trend culminated in 1952, when *Blue Poles: Number 11, 1952* was painted with a D_L value of 1.98. Typical results mapping out this evolution of D values with the years are summarized in Table 1. Also summarized in the table are the canvas surface area and the percentage of this area covered by the painted pattern (A).

In figure 9(a) the D_D values of typical paintings are plotted against the year in which they were painted. The dashed lines are included as guides to the eye and reveal the basic trend in the evolution of D_D. The graph displays a rapid increase in D_D during the evolution from Pollock's "preliminary" to "transitional" phase as he established his drip technique, followed by a more gradual increase as he refined his technique towards the "classic" style. His drip technique evolved considerably during the period 1943–1952. His initial drip paintings of 1943 consisted of a few dripped trajectories which, although distributed across the whole canvas, occupied less than 10% of the $0.36m^2$ canvas area. By 1952 he was spending six months laying down extremely dense patterns of trajectories which covered over 90% of his vast $10.22m^2$ canvas. Figure 9(b) and (c) shows a correlation between the high D_D values of his "classic" patterns and his use of a large canvas and high pattern density.

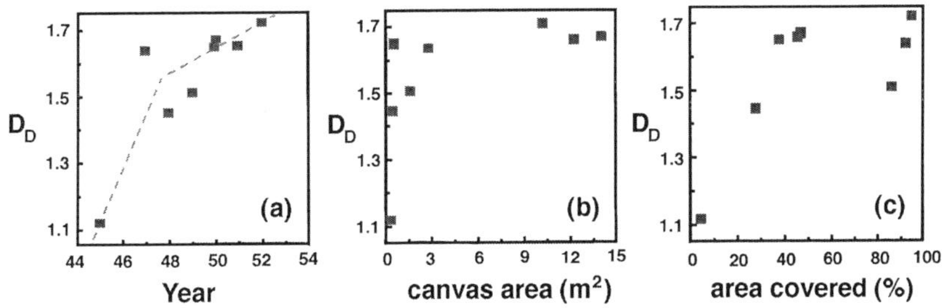

FIGURE 9 The Evolution of Pollock's paintings through the years. The painting's fractal dimension D_D is plotted against (a) the year in which it was painted, (b) the canvas area and (c) the percentage of this area covered by paint. In (a) the dashed line is a guide to the eye.

5 THE FRACTAL CONSTRUCTION PROCESS

How did Pollock construct and refine his fractal patterns? In many paintings, though not all, Pollock introduced the different colors more or less sequentially: the majority of trajectories with the same color were deposited during the same period in the painting's evolution. To investigate how Pollock built his fractal patterns, I have electronically de-constructed the paintings into their constituent colored layers and examined the fractal content of each layer. The analysis shows that each individual layer consists of a uniform fractal pattern. The initial layer in a Pollock painting plays a pivotal role within the multilayer construction—it has a significantly higher fractal dimension than subsequent layers. This layer essentially determines the fractal nature of the overall painting, forming the foundation of the painting and acting as an "anchor layer" for the subsequent layers. The black anchor layer of the painting *Autumn Rhythm: Number 30, 1950* is shown in color plate 14. As subsequent layers are added to this painting, the D_D value rises only slightly—from 1.66 (with just the black anchor layer) to 1.67 (the complete painting with all the layers). In this sense, the subsequent layers merely fine tune the D_D value established by the anchor layer. The anchor layer also visually dominates the painting. Pollock often chose the anchor layer to be black, which contrasts against the light canvas background. Furthermore, the anchor layer occupies a larger surface area than any of the other layers. For *Autumn Rhythm: Number 30, 1950* the anchor layer occupies 32% of the canvas space while the combination of the other layers—brown, gray, and white—occupies only 13% (the remaining 55% corresponds to exposed canvas).

Since the fractal content and visual character of a Pollock painting is determined predominantly by the anchor layer, I will examine the evolution of this layer in detail. In the anchor layer's initial stage, the trajectories are grouped into small, unconnected "islands," each of which is localized around a specific region of the canvas. Pollock then went on to paint longer trajectories. These extended trajectories joined the islands, gradually submerging them in a dense pattern of trajectories which became increasingly fractal in character. The visual evolution of

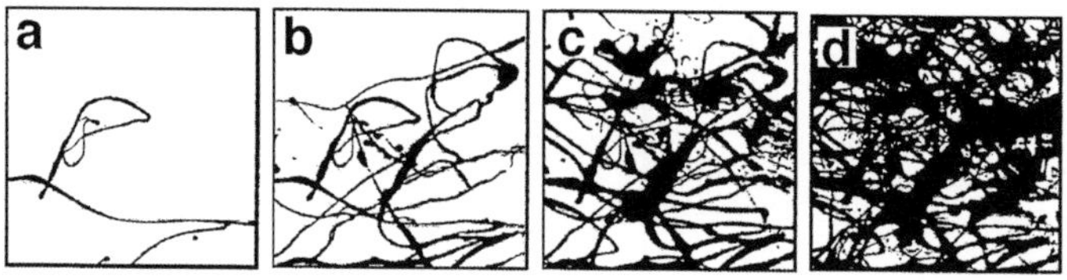

FIGURE 10 A 69cm by 69cm section of a painting filmed (a) 5, (b) 20, (c) 27, and (d) 47 seconds into Pollock's painting process. The film was shot by Hans Namuth in 1950.

this process is documented in figure 10 which shows a processed image of a painting filmed during Pollock's "classic" period of 1950.[9] Pollock painted the image on a glass surface with the camera recording from below. Using a video recording of the original film, the image was processed to remove background color variations. The image also was reflected around the vertical axis so that the final image appears from Pollock's point of view. The resulting black and white representation shown in figure 10 was then converted into a bitmap format for fractal analysis. The film records a 69cm by 69cm region lying within one of the islands. To analyze the fractal content the computer covers this region with a mesh of identical squares as described earlier. The first image, shown in figure 10(a), is recorded at 5 seconds into the painting process. At this initial stage of the painting, the scaling plot fails to condense onto a straight line—indicating that painting is not fractal at this early stage. As Pollock starts to paint more extended trajectories, the pattern density of the painting starts to rise rapidly with time. This rise in pattern density with time T is quantified in Table 2, where the percentage of the canvas area occupied by the painted pattern, A, is shown to rise rapidly over the first minute—by $T = 47$s, more than two thirds of the surface is covered with paint. Figure 10(a)–(d) show that the rapid rise in A with time is accompanied by an increase in spatial uniformity. The scaling plots confirm the introduction of fractal content. By $T = 27$s, the scaling plot condenses onto a straight line and a value of $D = 1.72$ is obtained from the gradient. By $T = 47$s, D has risen to 1.89, reflecting the rich complexity of fractal structure in the pattern shown in figure 10(d). At this stage, after less than one minute, the crucial stage of Pollock's fractal generation process is over: the anchor layer has been defined.

Labeling the formation of the anchor layer as phase one, and the subsequent multilayer fine-tuning process as phase two, for some of Pollock's works there was also a phase three, which took place after the painting process was completed. The fractal character of the completed patterns sometimes deteriorated towards the canvas edge. To compensate for this Pollock cropped some of his canvases after he had finished painting, removing the outer regions of the canvas and retaining the highly fractal central regions. The complete paintings, generated by this highly systematic three-phase process, follow the fractal scaling relationship (the straight line within the scaling plots) with remarkable accuracy and consistency. How did

[9] This section of a painting was filmed by P. Falkenberg and H. Namuth. The completed painting, which covered 121.9cm by 182.9cm, no longer exists [11].

TABLE 2 A summary of the anchor layer's parameters as they evolve during the first 47 seconds of the painting process.

T(s)	D_D	A (%)
5	–	3.3
20	1.52	16.5
27	1.72	42.5
47	1.89	70.2

Pollock arrive at this remarkable fractal generation process? Some insight can be obtained by considering investigations of human aesthetic judgements of fractal images. A recent survey revealed that over ninety percent of subjects found fractal imagery to be more visually appealing than non-fractal imagery and it was suggested that this choice was based on a fundamental appreciation arising from humanity's exposure to Nature's fractal patterns [28]. The survey highlighted the possibility that the enduring popularity of Pollock's Fractal Expressionism is based on an instinctive appreciation for Nature's fractals shared by Pollock and his audience.

It is clear from the analysis that Pollock's painting process was geared to more than simply generating a fractal painting—if this were the case he could have stopped after twenty seconds (for example the image in figure 10(b) is already fractal). Instead he continued beyond this stage and used the three-phase process over a period lasting up to six months. The result was to fine tune the patterns and produce a fractal painting described by a highly specific D_D value. The investigations shown in figure 9(a) show that Pollock refined his technique through the years, with the D_D value of his completed paintings rising from 1.12 in his early attempts in 1945 to 1.72 at his peak in 1952. In 1950 Pollock generated a fractal painting characterized by $D_D = 1.89$ (see fig. 10(d)). However, he immediately rubbed out this pattern and started again, suggesting that a pattern with such a high fractal dimension wasn't visually appealing to him. The highest D_D of any completed painting is for *Blue Poles: Number 11, 1952* with a value of 1.72. This was painted towards the end of his career in 1952. Therefore, it can be speculated that Pollock's quest was to paint drip patterns characterized by approximately $D_D = 1.7$. Why would Pollock refine his process to generate fractals with high D_D values? It is interesting to note that, in a recent survey designed to investigate the relationship between a fractal pattern's D value and its aesthetic appeal, subjects expressed a preference for patterns with D values of 1.8 [16], similar to Pollock's "classic" paintings of 1950–1952. Although a subsequent survey reported much lower preferred values of 1.26, this second survey indicated that self-reported creative individuals have a preference for higher D values [2], perhaps compatible with Pollock's quest to paint patterns with high D values. Table 3 lists the D values for examples of Nature's scenery. It is interesting to note that Pollock's "preferred" D value corresponds to the fractal dimension of a scenery familiar to us all in our every day lives—trees in a forest. It is possible to speculate therefore

TABLE 3 Common natural scenery and their fractal dimensions.

Scenery	D
Cauliflower	1.1 – 1.2
Mountain Profile	1.2
Stars	1.2
Coastlines	1.2 – 1.3
River	1.2 – 1.3
Waves	1.3
Lightning	1.3
Volcanic Cloud	1.3
Clouds	1.3
Mud Cracks	1.7
Ferns	1.8
Forests	1.9

that Pollock's paintings were an expression of a fundamental appreciation of the natural scenery which surrounded him—an appreciation acquired either through evolution or learnt implicitly through his life. Within this context it is significant to note that Pollock's development of the drip and splash technique occurred as he moved from downtown Manhattan (artificial scenery) to the countryside (natural scenery). Perception studies are planned to examine these possibilities further.

In addition to exploring the aesthetic appeal of Pollock's patterns, perception studies also may provide an answer to one of the more controversial issues surrounding Pollock's drip work. Over the last fifty years there has been a persistent theory which speculates that Pollock painted illustrations of objects (for example, figures) during early stages of the painting's evolution and then obscured them with subsequent layers of dripped paint [11, 31]. Since fractal patterns do not incorporate any form of figurative imagery, my analysis excludes the possibility that the initial stages of his paintings featured painted figures. Why, then, is the "figurative" theory so persistent? A possible answer can be found by considering my analysis in the context of the perception studies of Rogowitz and Voss [21]. These studies indicate that people perceive imaginary objects (such as human figures, faces, animals etc.) in fractal patterns with low D values. For fractal patterns with increasingly high D values this perception falls off markedly. Rogowitz and Voss speculate that their findings explain why people perceive images in the ink blot psychology tests first used by Rorschach in 1921. Their analysis shows that ink blots are fractal with a D value close to 1.25 and thus will trigger perceptions of objects within the patterns. Although not discussed by the authors, their results may explain the Surrealist method of Free Association where the artist stares at painted patterns until an image appears [1]. It could be that the patterns produced by the Surrealists (e.g., the Ernst's "frottage," Dominguez's "decalcomania," and Miró's washes) were fractal patterns of low dimension. Their findings also explain why figures are perceived in the initial layers of Pollock's paintings. The fractal analysis of the evolution of Pollock's patterns shows that his paintings started with a low D value which then gradually rose in value as the painting evolved towards completion (see table 2). Thus it is consistent with the findings of Voss and Rogowitz that an observer would perceive objects in the initial patterns of a Pollock painting (even though they are not there) and that these objects would "disappear" as D rose to the high value which characterized the complete pattern.

6 FRACTAL ANALYSIS AND JUDGEMENTS OF AUTHENTICITY

Finally, I briefly will consider the use of the fractal analysis technique to authenticate a Pollock drip painting. The results presented so far emphasize that fractal patterns are not an inevitable consequence of dripping paint—it is possible to generate drip paintings which have a non-fractal composition (see, for example, figure 4 (top)). Indeed, to emphasize this fact I have analyzed the paint marks found on the floor of Pollock's studio. Although dripped, these patterns are not fractal. In contrast, the patterns on Pollock's canvases are the product of a specific drip technique engineered to produce fractals and all of the drip paintings I have analyzed (over 20 in number) have this fractal composition. Therefore, fractality can be identified as the "hand" of Pollock and a fractal analysis can be used to authenticate a Pollock drip painting. Developing this argument one step further, the analysis also may be used to date an authentic Pollock painting. The D_D value of Pollock's work rose through the years, following a predictable trend (see for example figure 9(a)). Charting this progress, it should be possible to determine the D_D value of the painting and from this to suggest an approximate date. Recently, these proposals were put to the test. Color plate 15 shows a drip painting of unknown origin which lacks a signature but is thought to have been painted during Pollock's era. The painting was sent to me by a private collector in the USA to determine if the painted patterns were consistent with Pollock's. Painted using a drip technique, the painting has a uniform quality lacking any center of focus. These characteristics are shared with Pollock's "all-over" style. However, unlike a typical Pollock painting, my analysis shows that patterns at different magnifications are not described by the same statistics—the scaling plot fails to condense onto a straight line. It is not possible to characterize this painting with a D_D or D_L value and to plot it along side Pollock's drip paintings in figure 9. Despite any superficial similarities with Pollock's work, this painting does not contain fractal patterns—the characteristic "hand" of Pollock is absent. Clearly this use of a computer to detect the fundamental characteristics of painted patterns is a powerful one and will become part of a growing collection of scientific tools (which already includes techniques such as X-ray analysis to detect patterns hidden underneath subsequent layers of paint) employed by art theoreticians to investigate works of art. Such developments are a signal of the growing interplay between art and science.

7 CONCLUSIONS

The profound nature of Pollock's contribution to modern art lies, not only in the fact that he could paint fractals on a canvas, but in how and why he did so. In this chapter I have used a fractal analysis technique to examine the painting process Pollock used to construct his drip paintings. This analysis reveals a remarkably systematic method capable of generating intricate patterns which obey the fractal scaling behavior with precision and consistency. These results have been presented within the context of recent perception studies of the aesthetic appeal of fractal patterns. Nature builds its patterns using fractals as its basic building block. Having evolved surrounded by this fractal scenery, it perhaps therefore is not surprising that humanity possesses an affinity with these fractals and an implicit recognition of their qualities. Indeed, it is possible to speculate that people possess some sort of "fractal encoding" within the perception systems of their minds. The study of human responses to fractal images and the characterization of their aesthetic appeal is a novel field of research for perceptual psychologists, one which offers huge potential [2, 4, 6, 12, 16, 19, 21, 28]. Pollock's enduring popularity may be a consequence of a shared appreciation of Nature's fractal patterns operating within the psyche of both the painter and the observer. This chapter therefore raises a fundamental question: could Pollock have distilled Nature's very essence—fractal patterning—from within his mind and recorded this imagery directly on canvas? As Pollock himself noted, "Painting is self-discovery. Every good painter paints what he is,"[10] concluding that "I am nature."[11] Clearly, a discussion of Pollock's fractals would be incomplete without considering the art historical context of his work. It is hoped, therefore, that the results presented here will stimulate a debate between scientists, psychologists and art theoreticians regarding the artistic significance of Pollock's fractal drip paintings.

ACKNOWLEDGMENTS

Adam Micolich and David Jonas for their invaluable input during the development of the fractal analysis techniques. Michele Taylor for many useful debates on Pollock's fractals. James Coddington, Chief Conservator at the Museum of Modern Art, New York, for useful discussions on the potential of the analysis for authentification of Pollock's paintings. Glenn Day for his permission to analyze and show the drip painting shown in color plate 15.

REFERENCES

[1] Ades, D. "Dada And Surrealism." *Concepts of Modern Art*, edited by Nikos Stangos. London: Thames and Hudson, 1974.

[2] Aks, D. J., and J. C. Sprott. "Quantifying Aesthetic Preference For Chaotic Patterns." *Empirical Studies Of The Arts* **14** (1996): 1.

[10] O'Conner [14, p.226].

[11] O'Conner [14, p.226]

[3] Avnir, D., O. Biham, D. Lidar, and O. Malcai, "Is The Geometry Of Nature Fractal?." *Science* **279** (1998): 39–40.
[4] Briggs, J. *Fractals—The Patterns Of Chaos.* London: Thames and Hudson, 1992.
[5] Friedman, B. H. *Jackson Pollock: Energy Made Visible*, 93. New York: McGraw-Hill, 1972.
[6] Geike, J., and G. Landini, "Individual Differences In The Perception of Fractal Curves." *Fractals* **5** (1997): 129–143.
[7] Gleick, J. *Chaos* New York: Penguin Books, 1987.
[8] Gouyet, J. P. *Physics And Fractal Structures.* New York: Springer-Verlag, 1996.
[9] Klafter, J., M. F. Shesinger, and G. Zumofen. "Beyond Brownian Motion." *Physics Today* **49** (1996): 33–39.
[10] Knill, D. C., D. J. Field, and D. Kersten, "Human Discrimination of Fractal Images." *J. Opt. Soc. Amer.* **76** (1990): 1113–1123.
[11] Landau, E. G. *Jackson Pollock.* London: Thames and Hudson, 1989.
[12] Mandelbrot, B. B. *The Fractal Geometry Of Nature.* New York: W.H. Freeman and Company, 1977.
[13] Namuth, Hans. "Photographing Pollock." In *Pollock Painting*, edited by Barbara Rose, 89. New York: Agrinde Publications, 1980.
[14] O'Conner, Francis V., and Eugene Victor Thaw. *Jackson Pollock: A Catalogue Raisonné of Paintings, Drawings And Other Works*, vol 4, 248–251, New Haven, CT: Yale University Press, 1951.
[15] Ott, E. *Chaos In Dynamical Systems.* Cambridge, Cambridge University Press, 1993.
[16] Pickover, C. A. *Keys To Infinity.* New York: Wiley, 1995.
[17] Pickover, C. A. *Patterns, Chaos And Beauty: Graphics From The Unseen World.* New York: St. Martins, 1990.
[18] Pietgen, H. O. *The Art Of Fractals. A Computer Graphical Introduction.* Berlin: Springer-Verlag, 1988.
[19] Rodman, Selden *Conversations With Artists*, 82. New York: Devin-Adair, 1957.
[20] Pollock, Jackson."My Painting." *Possibilities* **1** (1947): 79.
[21] Rogowitz, R. E., and R.F. Voss, "Shape Perception And Low Dimension Fractal Boundary Contours." *Proceedings Of The Conference On Human Vision: Methods, Models And Applications*, edited by B. E. Rogowitz and J. Allenbach, 387–394. S.P.I.E. 1249 1990.
[22] Rosenberg, Harold. "The American Action Painters." *Artnews* **51** (1952): 22–23.
[23] Rubin, William "Jackson Pollock And The Modern Tradition: Part One." *Artforum* **5** (1967): 19.
[24] Shaw, R. *The Dripping Faucet As A Model Chaotic System.* Santa Cruz, CA: Aerial Press, 1984.
[25] Taylor, R. P. "Jackson Pollock: Nature, Chaos and Fractals." Master of Art Theory thesis, University of New South Wales, Sydney, 2000.
[26] Taylor, R. P., A. P. Micolich, and D. Jonas. "Fractal Analysis Of Pollock's Drip Paintings." *Nature* **399** (1999): 422.

[27] Taylor, R. P., A. P. Micolich, and D. Jonas, "Fractal Expressionism." *Physics World* **12** (1999): 25.
[28] Taylor, R. P. "Splashdown." *New Scientist* **159** (1998): 30–31.
[29] Tritton, D. "Ordered And Chaotic Motion Of A Forced Spherical Pendulum." *European J. Phys.* **7** (1986): 162.
[30] Tsallis, C. "Lévy Distributions." *Physics World* **10** (1997): 43–45.
[31] Varnedoe, K., and P. Karmel. *Jackson Pollock.* New York: Abrams, 1998.
[32] Voss, R. F. "Local Connected Fractal Dimension Analysis Of Early Chinese Landscape Paintings And X-Ray Mammograms." In *Fractal Image Coding And Analysis*, edited by Y. Fisher. NATO ASI Series F. Norway: Springer-Verlag, 1995.

Art and Complexity
J. Casti and A. Karlqvist (editors)

A Depth-First Traversal of Thinking

Gail Wight

How thinking happens, why it happens, how much of our thinking is simply a side-effect, how fragile our thoughts might be, how the definition of thinking and its importance changes with time—these are some of the questions that infect my art-making. The pieces described below play with the subject of thinking, and I've broken them down into three interrelated categories: animal cognition, machine intelligence, and scientific pedagogy. The third category informs the first two, since it's the lens through which they come into focus.

For instance, current theories in complexity science have enabled us to consider emergent properties as a possible model for the how thinking happens—how the sum could become more than its individual parts. Emergent properties are also being pursued as the key to creating an artificial intelligence, as machines begin to amass experience in various learning processes. In turn, ideas about emergent properties have suggested new ways for artists to approach art-making. While I hope the following categories invite a brief depth-first traversal of thinking, the truth is that the influence of scientific pedagogy permeates throughout.

1 ANIMAL COGNITION

ONE HUNDRED LINKS

"One Hundred Links" is a response to Rousseau's eighteenth century critique of the chains that bind us to civilization. In the late-twentieth century, our own disheveled neurochemistry would be identified as the likely culprit for binding us to social mores. To break those chains, visitors are invited to sample one hundred potential states of mind. Neurotransmitters in solution are labeled with the states they

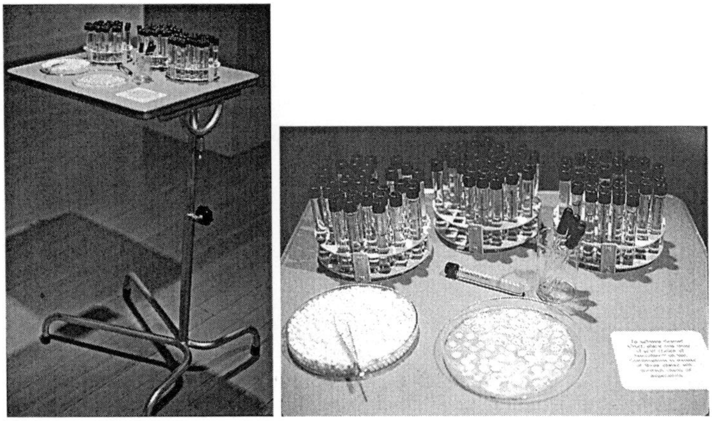

FIGURE 1 One Hundred Links (for Rousseau).

might effect. Those states are based on medical findings, philosophy, literature, and folklore.

HEREDITARY ALLEGORIES: A STUDY IN GENETICS

A residency at San Francisco's Capp Street Project enabled me to research the basic concepts and historical landmarks that define the study of genetics. These essential ideas and discoveries were presented as anecdotes, each with an added twist to encourage the viewer to question the underlying conventions and assumptions of this scientific and cultural phenomenon.

The stories were illustrated by thirty mice and a canary. As an example, take the case study of separated identical twins who had planted the same tree in their yards, and then built the same circular bench around the trees with no knowledge of the other's actions. This was illustrated by two mice living in separate cages, each with their own tiny pine trees circled by tiny wooden benches. They both ate away at their props throughout the exhibition, presumably with no encouragement from each other.

THE FIRST EVOLUTIONARY OCCURRENCE OF PAIN

Research into the history of pain suggests that the common land snail was the first animal to develop pain receptors. In this piece, blueprints of the snail's nervous system are laced with copper. The copper carries electricity to miniature scenes from human existence powering street lights and sounds, as a means to illuminate and amplify our inherited ability to sense pain. The inheritance of pain, however, grows infinitely complex within the realm of human experience. The scenes depicted are an impending assault, an unexplained car crash, and a desolate civic center.

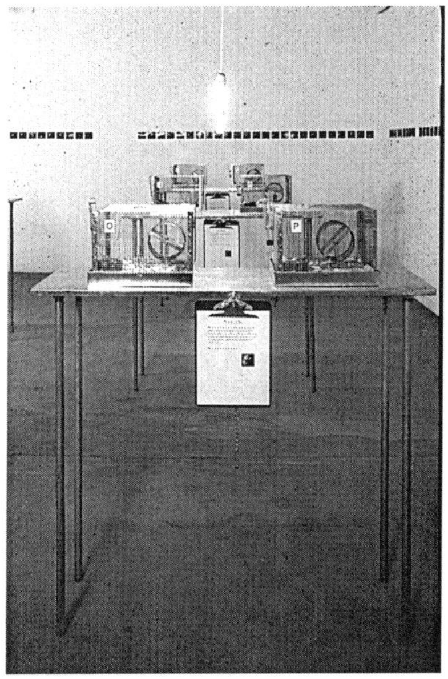

FIGURE 2 Hereditary Allegories: A Study in Genetics.

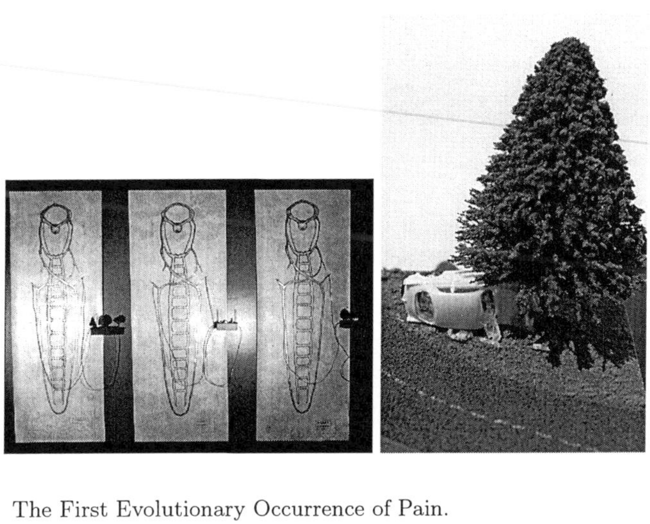

FIGURE 3 The First Evolutionary Occurrence of Pain.

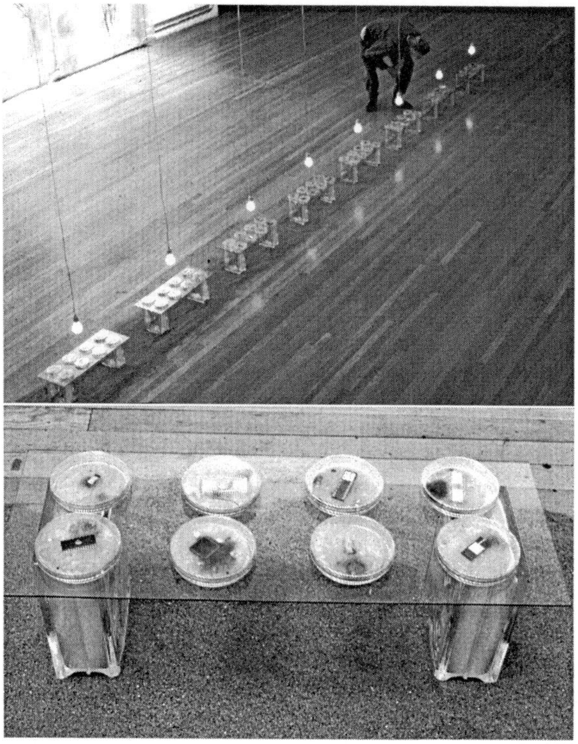

FIGURE 4 Residual Memory.

2 MACHINE INTELLIGENCE

RESIDUAL MEMORY

With fears and hyperbole mounting in popular culture about our future as cybernetic beings, embodying bits of machine technology, I wondered what the reverse perspective might be: do machines fear the encroachment of biology? I gathered sixty-four discarded computer chips from a Silicon Valley junkyard. I specifically chose central processing units (CPUs) and planted them in agar (a growth nutrient), to see if any biological lifeforms were invading these chips. It turned out that they were harboring all kinds of tiny flora and fauna, which began to flourish in the agar. These two forms of memory—one organic, one inorganic—created beautiful and fascinating combinations in their petri dishes.

CEREBRAL SONATA

The first human electroencephalogram ever recorded, along with a sampling of common EEGs ("brainwaves"), are converted into a stream of digits which, in turn, are played by a synthesized piano, violin, oboe, and organ. The history of

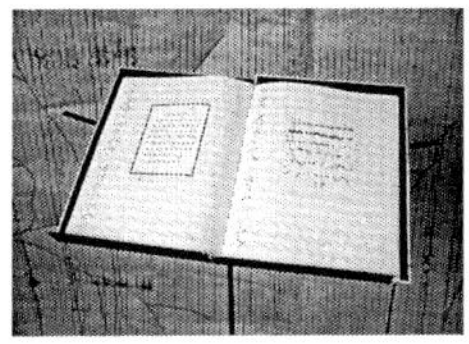

FIGURE 5 Cerebral Sonata.

reading cerebral electricity, from Neanderthals to present time, is presented on hand-made staff paper. The history ends with implications for shaping thought, as scientists learn to see these "brainwaves" in phase space. Housed in a tent made from contemporary EEG read-outs.

THE HISTORY OF WISHING

The archeological discovery of intentional Neanderthal burials, and the discovery that these burials included large amounts of wild flowers, is described in a series of museum cases. I hypothesize that this 60,000 year-old tradition of burying the dead with flowers must be hard-wired in the brain, and present diagrams to explain how this "wiring" might be physically organized. In a subtle injection of fiction, this now quantifiable act is burned onto a computer chip, suggesting that one could create an artificial intelligence that carries this trait. The final case contains the same species of flowers that were found in the archaeological dig—both real and artificial. Over the course of the exhibition, the real flowers wilt and die, leaving the artificial flowers in full bloom.

3 PEDAGOGY

NEURAL PRIMERS

In this set of five large books, each presents an animal that has a particular class of nervous system found in the animal kingdom. Each animal and its nervous structure has been chosen for a unique feature in the way it processes information, and for the resonant importance of that feature to humans, whether due to our similarity or dissimilarity. For instance, one book focuses on the octopus, which has three lobes rather than our two. The octopus is extremely emotional, changes colors due to emotional states, and can actually die from emotional shock.

A TALE OF TWO SLIMES

Commissioned by the Exploratorium for an exhibit of artworks based on the science of complexity, this book and time-lapse video tell the story of two slime molds. Both molds are important to the study of complexity for radically different reasons. The book examines these differences, and goes on to contrast their features, behaviors, and habitat, while the video presents time-lapse footage of each mold's

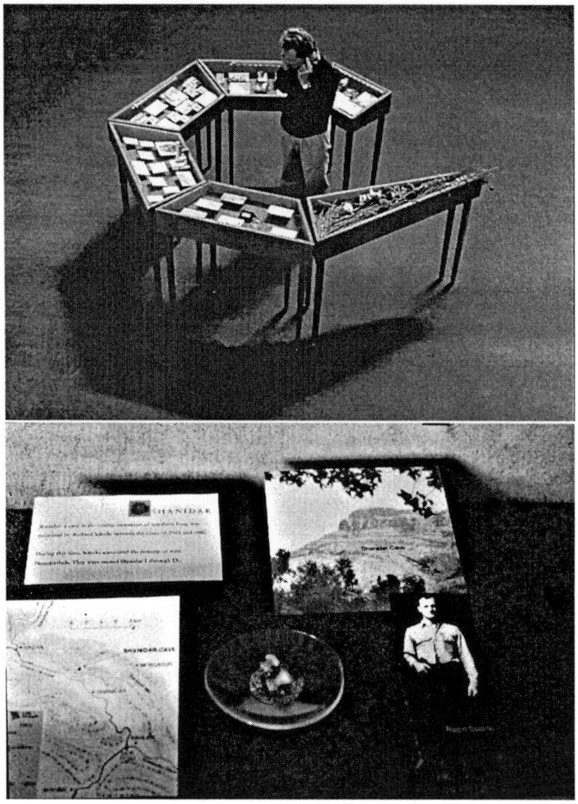

FIGURE 6 The History of Wishing.

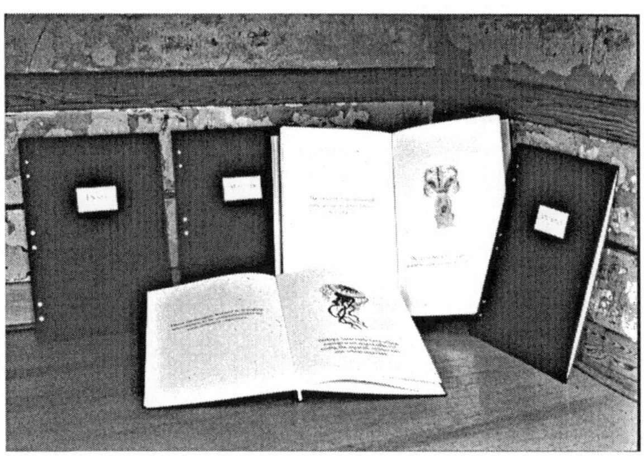

FIGURE 7 Neural Primers.

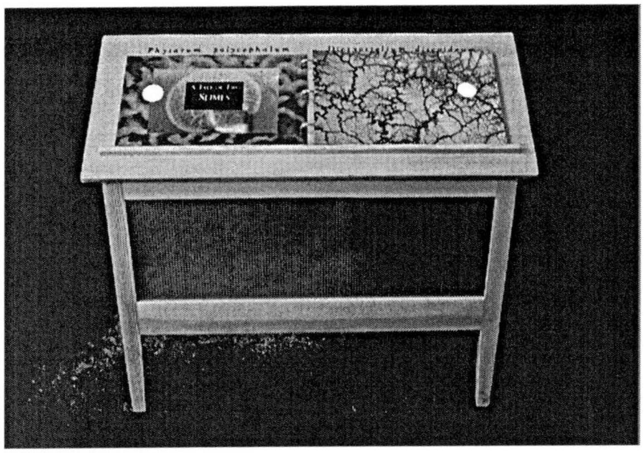

FIGURE 8 A Tale of Two Slimes.

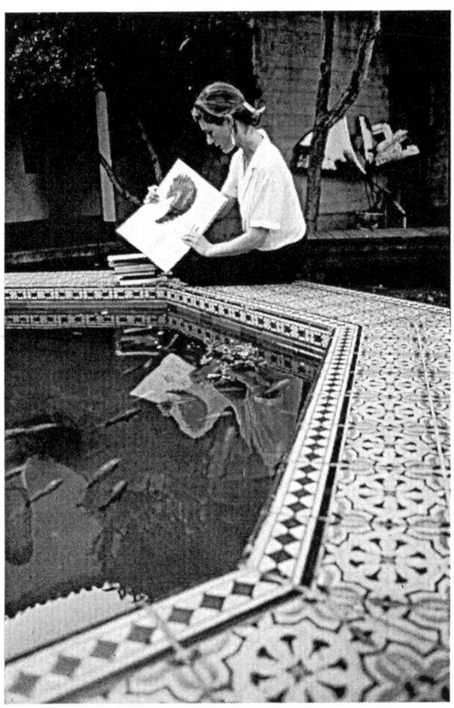

FIGURE 9 School of Evolution.

life-cycle. The book ends with an explanation of their only common trait, which binds all slime molds taxonomically: they leave behind a trace of their paths, a filmy residue of their travels.

SCHOOL OF EVOLUTION

A day long seminar was held for the fish at the San Francisco Art Institute. I scoured San Francisco's libraries for everything that science could tell fish about themselves, and then read to them for approximately six hours. Prehistory, genetics, anatomy and physiology of fish, as found in classic texts on ichthyology, were addressed. The day culminated with a special lecture on evolution and the slow formation of fins into legs. The fish were encouraged to follow their ancestors and put their minds to evolving out of the fish pond. I go back to check on them now and then.

Color Plates

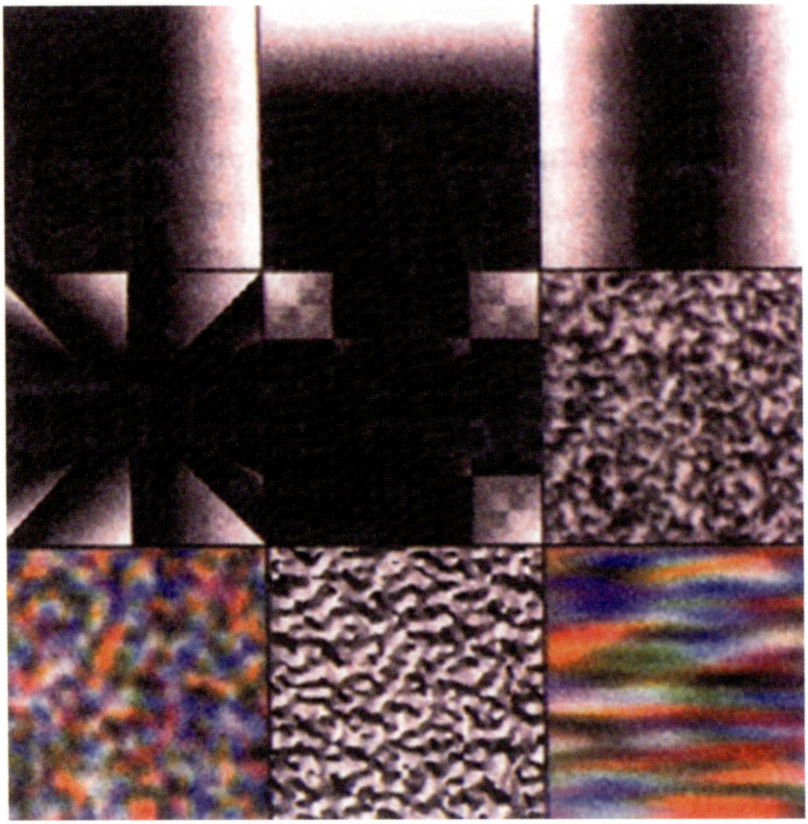

COLOR PLATE 1 The output from the simple LISP functions 1–9. (Casti, p. 21.) (Reprinted from Sims, K. "Artificial Evolution for Computer Graphics." *Computer Graphics* **25(4)** (1991): 319–328. ©, ACM Inc. by permission.)

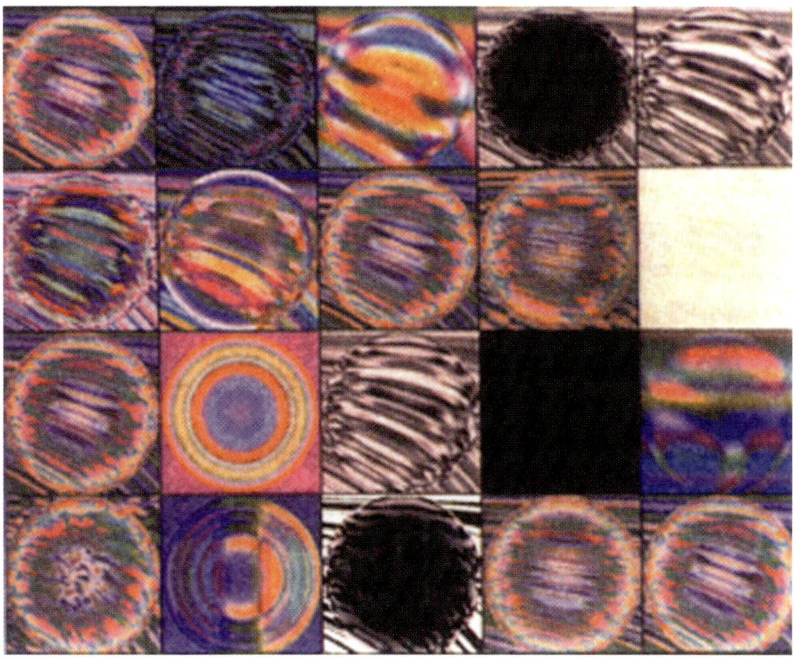

COLOR PLATE 2 A parent with 19 mutations. (Casti, p. 21.) (Reprinted from Sims, K. "Artificial Evolution for Computer Graphics." *Computer Graphics* **25(4)** (1991): 319–328. ©, ACM Inc. by permission.)

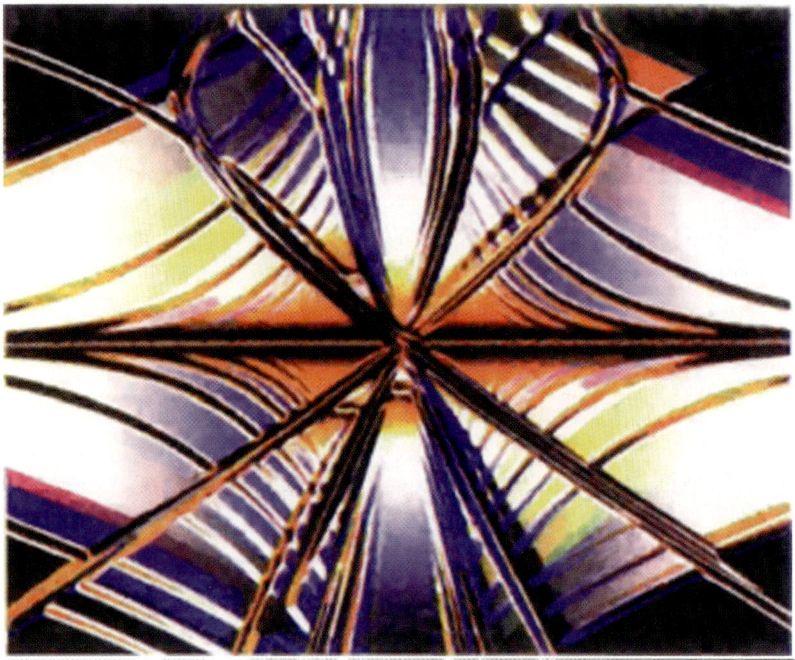

(round(log(*y (color-grad(round(*abs (round
(log(*y(color-grad(round(*y(log(invert y) 15.5))
x)3.1 1.86#(0.95 0.7 0.59) 1.35))0.19)x))(log
(invert y)15.5))x)3.1 1.9#(0.95(0.7 0.35)1.35))
0.19)x)

COLOR PLATE 3 (a) Evolved phenotypes and their corresponding genotypes. (Casti, p. 21.) (Reprinted from Sims, K. "Artificial Evolution for Computer Graphics." *Computer Graphics* **25(4)** (1991): 319–328. ©, ACM Inc. by permission.)

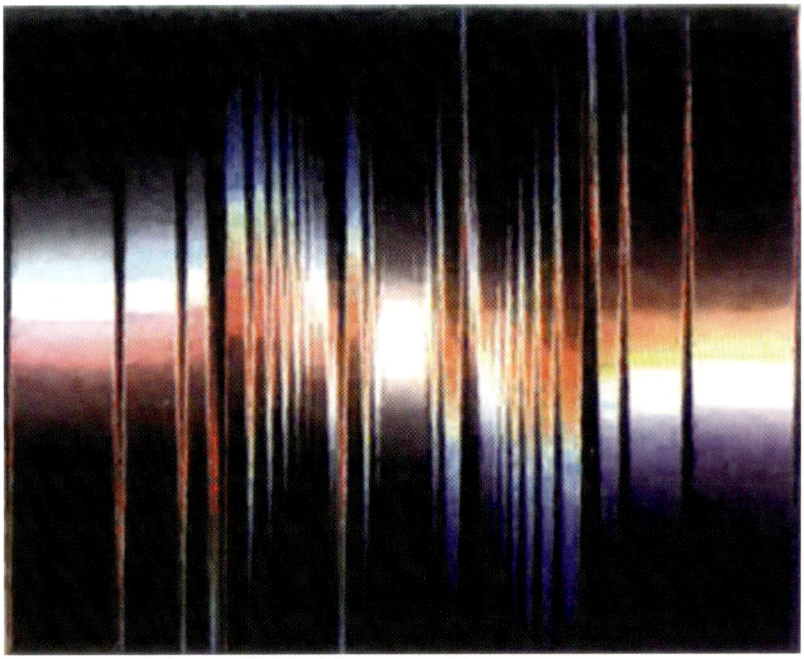

(rotate-vector(log(*y(color-grad(round(*(abs (round(log #(0.01 0.67 0.86)0.19) x))(hsv-to-rgb(bump(if x 10.7 y)#(0.94 0.01 0.4)0.78#(0.18 0.28 0.58)#(0.4 0.92 0.58)10.6 0.23 0.91)))x)3.1 1.93#(0.95 0.7 0.35)3.03))-0.03 x#(0.76 0.08 0.24))

COLOR PLATE 3 (b) Evolved phenotypes and their corresponding genotypes. (Casti, p. 21.) (Reprinted from Sims, K. "Artificial Evolution for Computer Graphics." *Computer Graphics* **25(4)** (1991): 319–328. ©, ACM Inc. by permission.)

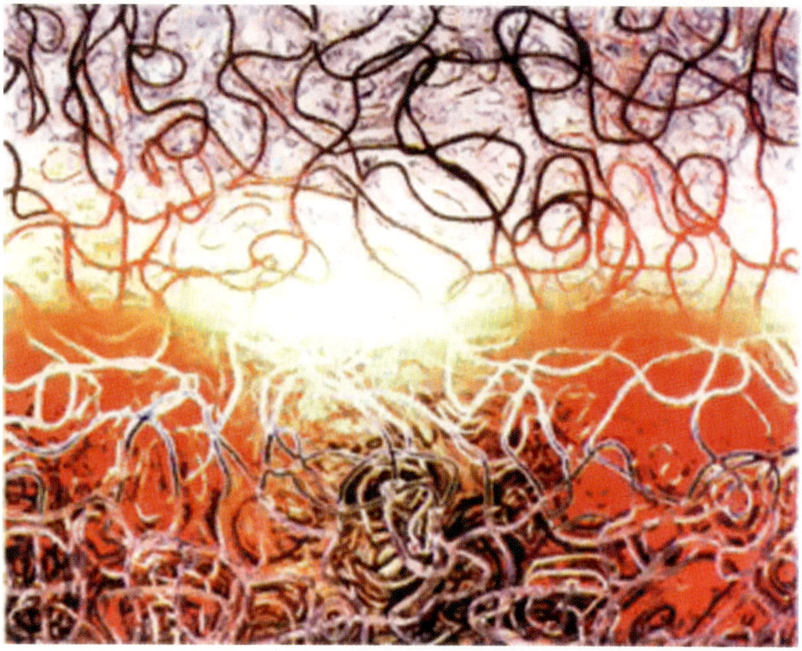

COLOR PLATE 3 (c) Evolved phenotypes and their corresponding genotypes. (Casti, p. 21.) (Reprinted from Sims, K. "Artificial Evolution for Computer Graphics." *Computer Graphics* **25(4)** (1991): 319–328. ©, ACM Inc. by permission.)

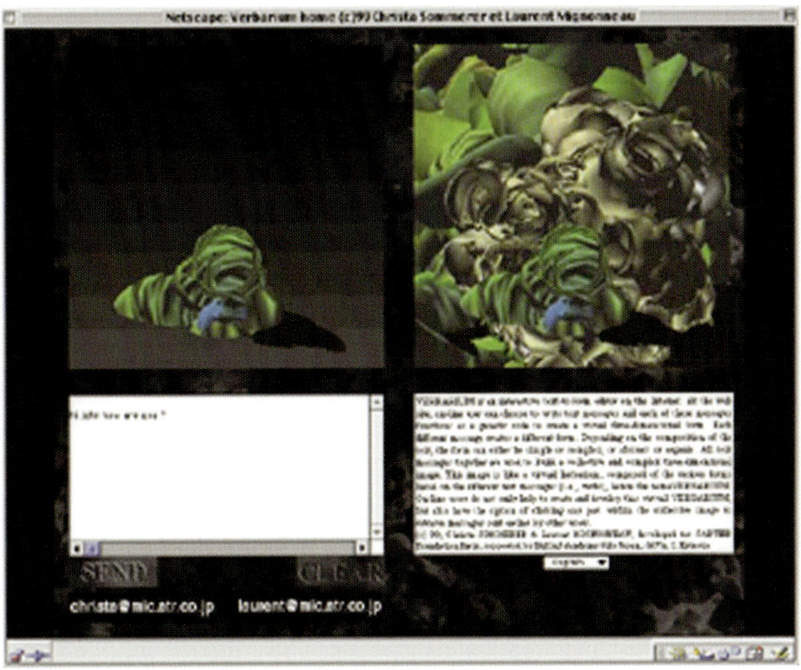

COLOR PLATE 4 VERBARIUM web page. (Sommerer, p. 85.)

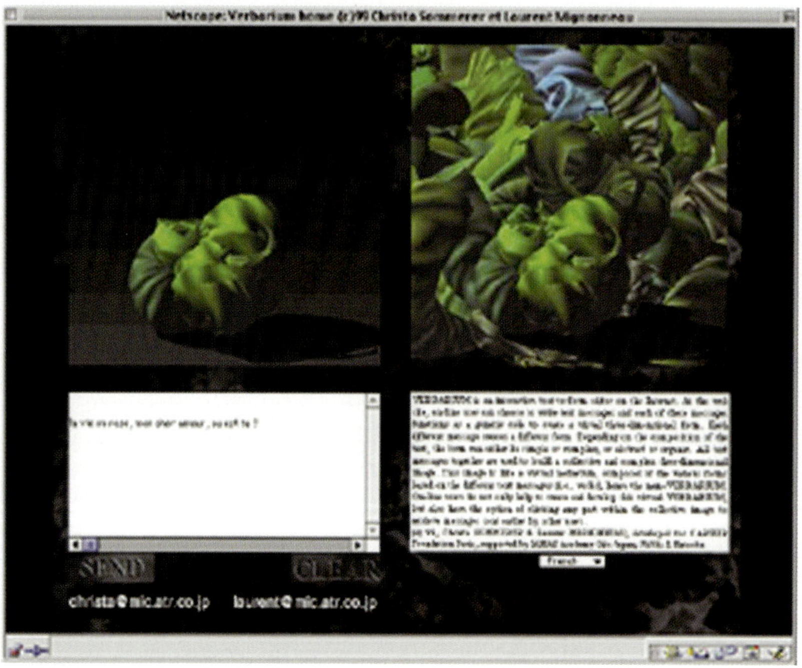

COLOR PLATE 5 VERBARIUM Web siteexample page. (Sommerer, p. 85.)

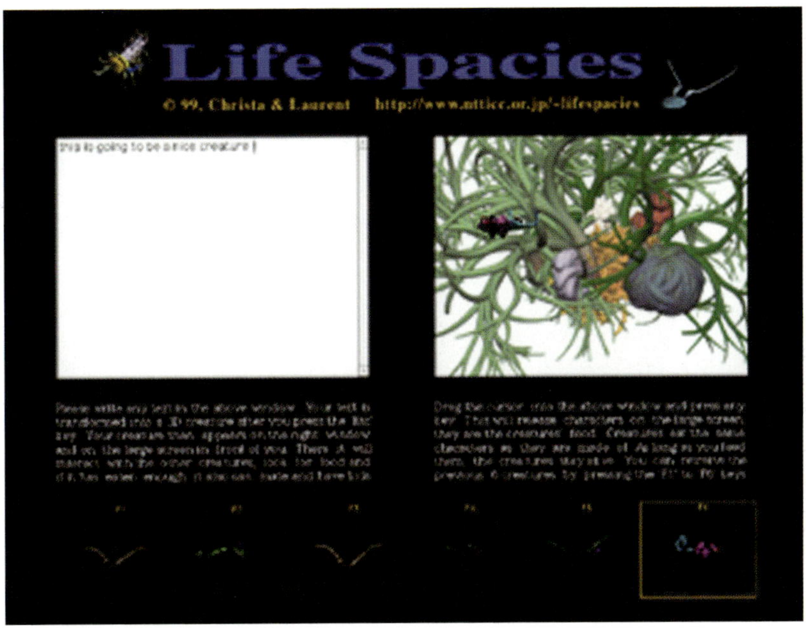

COLOR PLATE 6 Life Spacies IIgraphical user interface (GUI). The upper-left window is used to type messages and thus create creatures, and the upper-right window is used to place the cursor and release text characters to feed the creatures. (Sommerer, p. 85.)

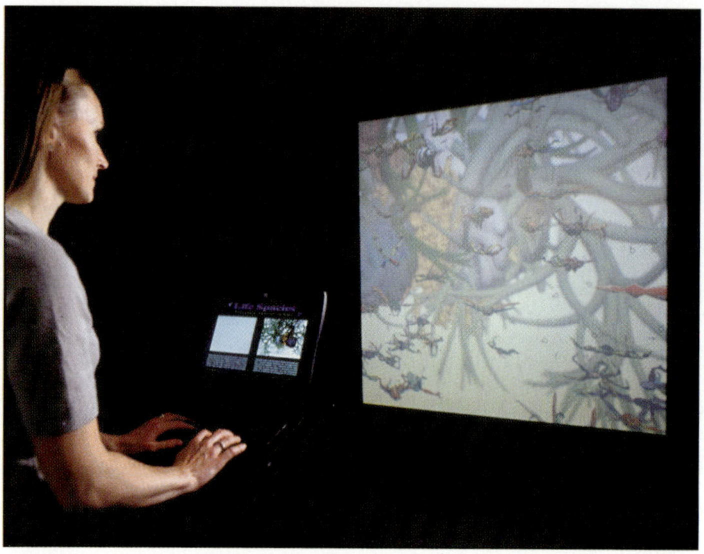

COLOR PLATE 7 "Life Spacies II"—user as she creates and feeds creatures on the GUI and watches them interact with other creatures on the large projection screen. (Sommerer, p. 85.)

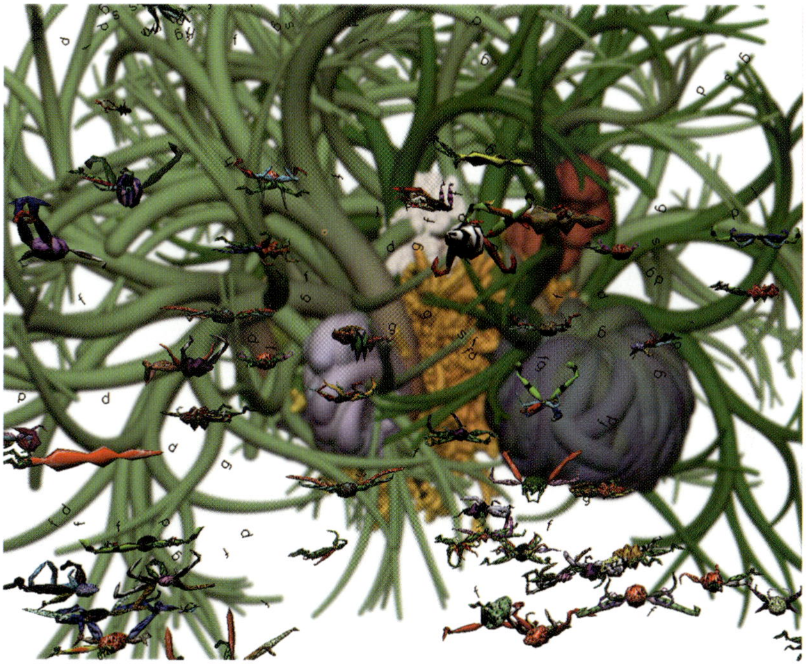

COLOR PLATE 8 Complex interaction among Life Spacies II creatures. (Sommerer, p. 85.)

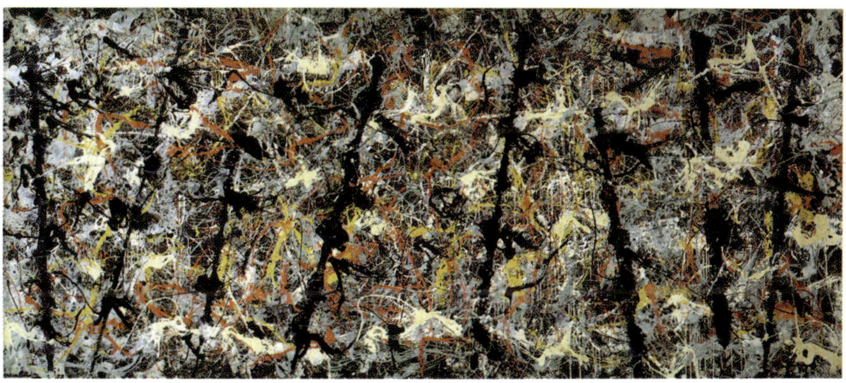

COLOR PLATE 9 Blue Poles: Number 11, 1952 (enamel and aluminum paint on canvas, 210cm by 486.8cm) was painted by Pollock in 1952 (The National Gallery of Australia, Canberra, Australia). (Taylor, p. 117.) (© 2002 The Pollock-Krasner Foundation/Artists Rights Society (ARS), New York.)

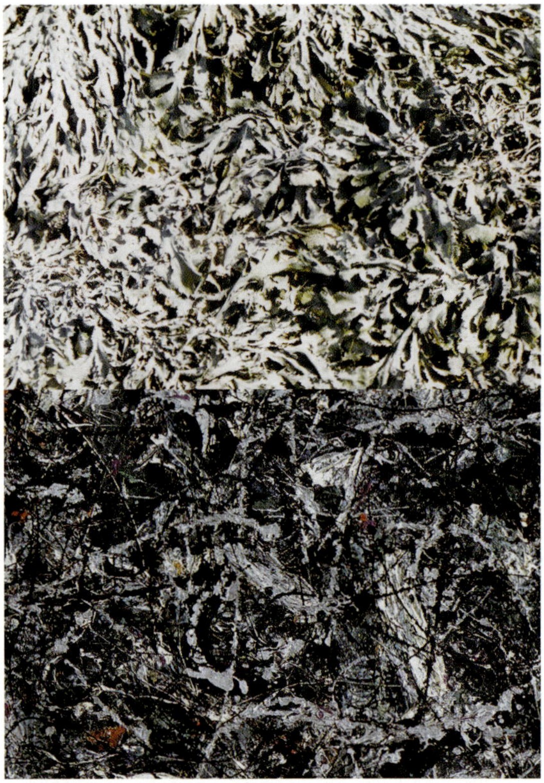

COLOR PLATE 10 A comparison of the patterns made by seaweed (top) and those within a 45.9cm by 69.6cm section of Full Fathom Five (bottom). Full Fathom Five (oil on canvas, 129.2cm by 76.5cm) was painted by Pollock in 1947 (The Museum of Modern Art, New York). The seaweed was photographed by R.P. Taylor. (Taylor, p. 117.)

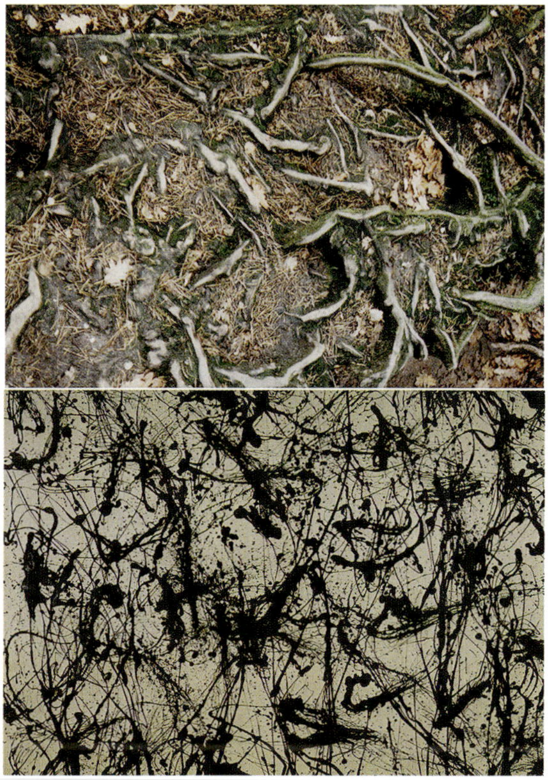

COLOR PLATE 11 A comparison of the patterns made by tree roots (top) and those within a 195.6cm by 294.1cm section of Pollock's painting Number 32, 1950 (bottom). The tree roots were photographed by R.P. Taylor. (Taylor, p. 117.) (© 2002 The Pollock-Krasner Foundation/Artists Rights Society (ARS), New York.)

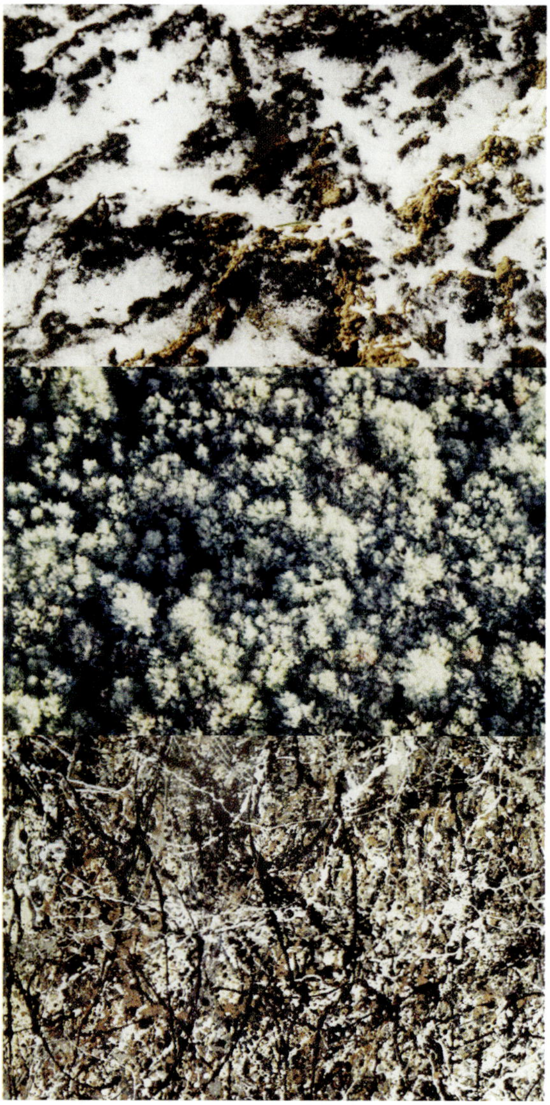

COLOR PLATE 12 Photographs of a 6.6cm by 10cm section of snow on the ground (top), a 3300cm by 5000cm section of forest (middle) and a 165cm by 250cm section of Pollock's One: Number 31, 1950 (bottom). The photographs of the snow and forest were taken by R.P. Taylor. The painting One: Number 31, 1950 (oil and enamel on canvas, 269.5cm by 530.8cm) was painted by Pollock in 1950 (The Museum of Modern Art, New York). (Taylor, p. 117.) (© 2002 The Pollock-Krasner Foundation/Artists Rights Society (ARS), New York.)

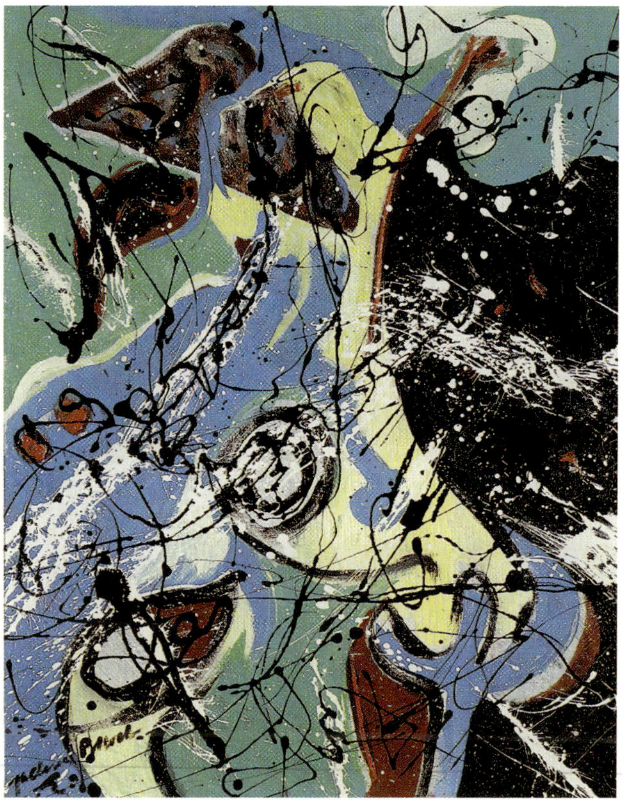

COLOR PLATE 13 Water Birds (oil on canvas, 66.4cm by 53.8cm) was painted by Jackson Pollock in 1943 (The Baltimore Museum of Art, Baltimore). (Taylor, p. 117.) (© 2002 The Pollock-Krasner Foundation/Artists Rights Society (ARS), New York.)

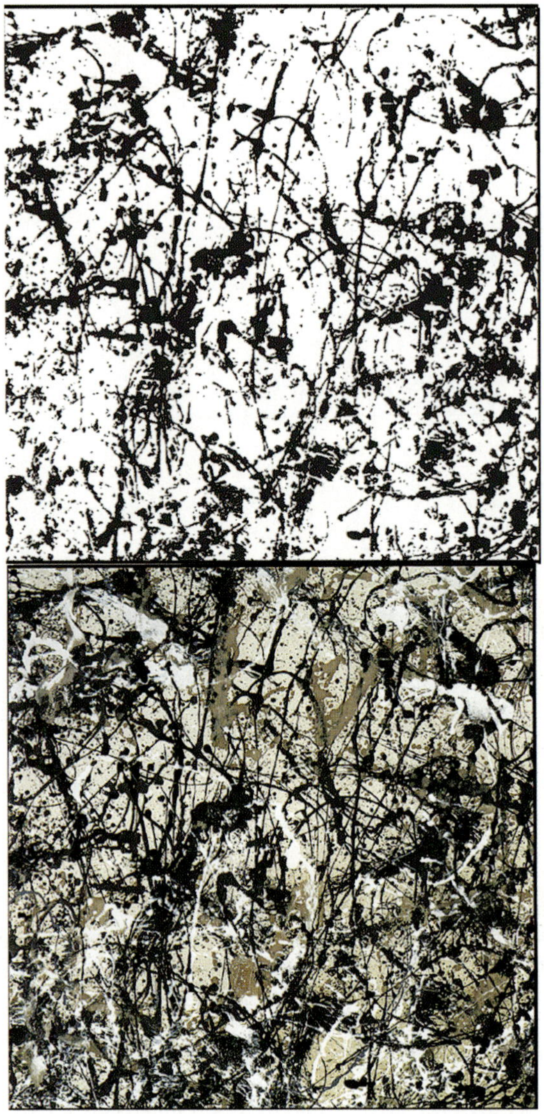

COLOR PLATE 14 A comparison of (top) the black anchor layer and (bottom) the completepattern consisting of four layers (black, brown, white and gray on a beige canvas) for the painting Autumn Rhythm: Number 30, 1950. Autumn Rhythm: Number 30, 1950 (oil on canvas, 266.7cm by 525.8cm) was painted by Pollock in 1950 (The Metropolitan Museum of Art, New York). (Taylor, p. 117.) (© 2002 The Pollock-Krasner Foundation/Artists Rights Society (ARS), New York.)

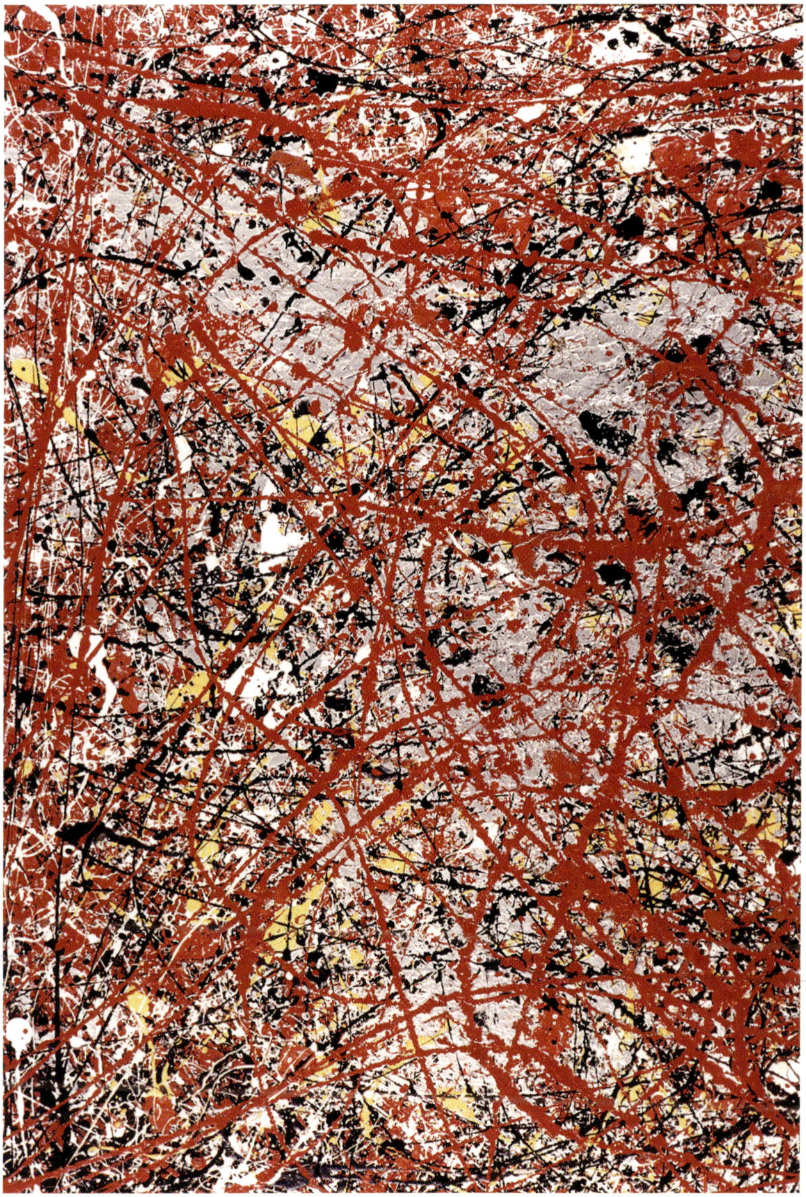

COLOR PLATE 15 A drip painting of unknown origin (61cm by 89cm) painted on masonite hard-board using aluminum, enamel and oil paints (Glenn Day, Texas). (Taylor, p. 117.)